Case Studies in Systems Analysis

Case Studies in Systems Analysis

J. P. A. Race

Senior Lecturer,
Department of Computer Science,
Brunel University

© J. P. A. Race 1979

All rights reserved. No part of this publication may be reproduced or transmitted, in any form or by any means, without permission.

First published 1979 by
THE MACMILLAN PRESS LTD
London and Basingstoke
Associated companies in Delhi Dublin
Hong Kong Johannesburg Lagos Melbourne
New York Singapore and Tokyo

Printed in Hong Kong

British Library Cataloguing in Publication Data

Race, John
   Case studies in systems analysis.
   1. Systems analysis - Case studies
   I. Title
   003   QA402

ISBN 0-333-23733-1

This book is sold subject to the standard conditions of the
Net Book Agreement.

The paperback edition of this book is sold subject to the condition that it shall not, by way of trade or otherwise, be lent, resold, hired out, or otherwise circulated without the publisher's prior consent in any form of binding or cover other than that in which it is published and without a similar condition including this condition being imposed on the subsequent purchaser.

Contents

1.	Introduction—Who Should Use this Book		1
2.	Advice to those Administering the Case Studies		4
3.	Advice to the Syndicate Members		7
	The Case Studies		
4.	Minims Restaurant	(B)*	12
5.	The Dental Supply Company	(C)	19
6.	Bluebird Aircraft	(B)	42
7.	The Garden Ornaments Management Information System Project	(A)	50
8.	Brighton Rock	(B)	63
9.	Camelot Ice Cream	(C)	66
10.	Vagabond Car Hire	(B)	74
11.	Canine Eugenics	(B)	77
12.	Banzai Pumps	(B)	81
13.	The PNP Electronics Company	(C)	85
14.	Hardware and Software for Case Studies		92
	Author's Notes on the Case Studies		97

*The case studies fall into the following categories:
(A) 2 hours reading and syndicate analysis followed by a half-hour presentation and discussion
(B) 5 hours syndicate work and tutorials followed by a written and/or verbal presentation and a one-hour de-briefing session
(C) 10 hours syndicate work and tutorials, otherwise as (B)

1
Introduction-Who Should Use this Book

Universities, polytechnics, schools, companies, and computer manufacturers and training organisations offer courses in <u>systems analysis</u>, by which we mean not only the study, but also the design and implementation of computer-based systems, to help companies and other institutions achieve their goals: typically profit and return on investment.

A 'system', in turn, can be defined in this context as

> 'an organisation of people with defined responsibilities using appropriate methods to achieve defined objectives.'

To train people in systems analysis in this sense involves the teacher in the following main topics. Firstly the tools of the analyst

- the hardware and software themselves, particularly the use of files - their structures and access methods

- techniques of analysing the present situation

- documenting systems using flow charts, block diagrams, decision tables, etc., as appropriate: report writing and presentation

- project evaluation using cost/benefit study, financial analysis, critical path methods, etc.

- project management: specifying, testing, documenting, training, installing, live running and post-installation audit.

Secondly the teacher will need to cover the main business application areas of computers

- Marketing: sales order processing, finished stock

control, sales ledger, statistics, etc.

- Production: assembly break-downs, net requirements planning, raw materials and work-in-progress control, progressing, etc.

- Administration: Research and Development, personnel, share register, corporate planning, etc.

In this second part of the course the students should be applying the techniques they learned in the first part. (A list of suitable textbooks for that is given in the bibliography.) But there are considerable difficulties in getting students to do this. Simply to describe a typical open-item sales ledger system will not bring home strongly enough the practical reasons why an open-item system is better - in some situations! - than a balance forward system, for maximising revenue, profit, and liquidity. Much more effective is the approach in which the students are given the problem of an imaginary company, and work out for themselves the solution which meets its special needs best.

Ideally the students will work in small syndicate teams charged with producing a report and/or a presentation on a case, and spending time together: individually, in syndicate, or in tutorial on the case, while at the same time attending formal lectures on the particular application area - marketing, production, or administration - in which their current case falls.

Besides being a nice technique for teaching, case study work is ideal for assessment. It does mean rather more effort has to be devoted to setting and marking than a traditional examination paper, but the teacher gets early feed-back in tutorial on both his own success in putting across the ideas, and that of the students in comprehending them.

Some aspects of a systems analyst's work are very hard to teach - the exercise of common sense, the ability to see that the problem as posed is the wrong one (and if so to re-negotiate the terms of reference to more appropriate ones) and a sensitivity to people. A case study should sometimes include some facts which will lead the syndicate to decide that experts other than computer professionals - management scientists, for example, or industrial relations specialists - should be brought in. The syndicates may sometimes be irritated by what they consider to be 'red herrings' - aspects of a case not relevant to their own defined roles - but they should reflect that in real life problems will be thrown at them which have not been carefully structured to be 'pure' computer problems, but will contain others which are the province of another specialist.

It would be wrong for us to encourage the systems professional to believe he can himself solve such problems, but he must be prepared to recognise them when they crop up during an assignment which is nominally a computer investigation, and not brush them to one side in order to concentrate on technical considerations. If this is done, the technical 'solution' will be no solution at all, and may be a gigantic irrelevancy.

In theory management should not send in a computer professional to investigate a problem which is partly or even entirely outside his field of competence: but this occurs, and when it does, the professional must recognise what is happening and react properly: separating out those aspects which are genuinely his own province, and recommending that the others be referred to the appropriate expert.

This is why these case studies are far from being 'chess problems' which have only one correct solution: they present, in many cases, a confused situation in which some facts are missing or even appear contradictory, and in which several courses of action are reasonable.

Maximum benefit is gained if the cases generate argument, frustration, and requests for clarification and further information: the teacher will find that using these cases may save some effort in inventing original situations, but he cannot sit back and let the syndicates get on with it - tutorials are essential, and it may be necessary to supply additional facts and figures, consistent with the original case, to satisfy the syndicates.

Accordingly the people who will use this book will be teachers of systems analysis giving the second half of such a course, whether in an academic or a business institution, and their students. The teachers should therefore have some practical experience themselves of the use of computers to solve real problems in industry and commerce.

2
Advice to those Administering the Case Studies

Forming Syndicates   Three to six individuals should form a team. If the group of students as a whole know one another fairly well, as they will if they have already done a 'techniques' part of the course already, it is sufficient to ask them to form themselves into syndicates and for them to give lists of members at the next lecture session. Even having a meal together beforehand may have prepared them to form spontaneously into syndicates.

Alternatively the course director can set up the syndicates arbitrarily, but a better spirit will arise if they form themselves.

On one occasion I had a very lazy student who was not wanted by any syndicate - until a particularly perceptive syndicate asked for him, because it realised that under the proposed assessment scheme it would actually benefit from a 'sleeping member'. The marks were allotted to individuals on the basis of their individual sections in their joint report: in addition, an overall mark for consistency was allocated and spread back among all the members. If one member wrote nothing at all, the others did not lose, and might in fact gain, since there was less chance of inconsistency in the final report if there was a section missing. This seemed to me to be an excellent example of true 'systems thinking' on the part of the syndicate!

Initial Briefing   The director should make the following clear.

- What the syndicate is to do (written and/or verbal presentation).

- What role the syndicate is to play, e.g. internal consultants, and to whom it is addressing its conclusions, e.g. the Managing Director. This may be stated in the case itself.

- When the report/presentation is to be given.

- What method of assessment is to be used. (It is a good practice to ask the syndicates how they would like to be assessed: in many cases the members prefer to have an overall mark distributed equally: or they may be allowed to state on their final report if they want this, or individual marks, or some compromise.)

- Arrangements for tutorials and the answering of specific questions. If there is a slightly competitive atmosphere among the syndicates, so much the better, but if one syndicate asks a perceptive question, it is wise to decide whether the answer must in fairness be broadcast to the other syndicates (for example, if an error has been found), or whether the syndicate should have the benefit of 'copyright' on its question. If you decide to tell the others, it helps to explain the reason to the original questioners.

- Roughly how much time the syndicate should spend. Ideally the syndicates will be mature enough to work out how much benefit in terms of knowledge (and marks!) they will get from the exercise, and allocate time, bearing in mind the need to do other things. Such cold-bloodedness is to be encouraged - that is systems thinking too.

Tutorials   The director will find that the best strategy is to let the syndicates read the case, and to have a full tutorial within a few days to explain queries. Thereafter, the director may simply offer appointments for individual syndicates or their representatives to come and see him. During lectures, the director can often use the case to illustrate his points. This always gets the attention of the audience, which is keen to pick up ideas. The lectures should if possible be organised to fit in with the likely sequence of thoughts going through the minds of the syndicates - for example, if they are deciding whether to introduce make-for-stock or to continue the present make-to-order policy, a couple of lectures on the implications of this decision will be very timely. Hence the planning of lectures should follow the choice of case study.

Assessment   The director may look at the report as a whole and use a marking scheme such as

Feasibility and technical competence	10
Analysis of present situation	10
Cost/benefit analysis	5
Security arrangements	5
Implementation schedule proposals	5
Relevance to needs of defined readers	5
Presentation	5
Consistency overall	5
	50

Alternatively he can mark each section of the report, typically

Introduction   (by Joe Bloggs)
Present situation and alternatives   (by Mary Smith)
Proposed new system, phase 1, etc.

and give up to 10% extra for overall consistency at the end.

Debriefing   It is essential that the reports do not 'disappear into a void' like examination scripts. In a university it is often not policy to discuss actual marks given, but at all events a rough indication should be given of the quality of each submission to its authors, and the director should tell the whole group of students about the different approaches that arose, and his own views. While the submissions are being assessed it helps if a note book is kept, in which to jot down such points of general interest.

I have written some notes on each case in section 6. These have been kept brief so that students and directing staff will use them more as guidelines and thought-provokers than as model answers. In many cases it may well be that there are other, and better, approaches than the ones I have suggested.

3
Advice to the Syndicate Members

<u>Roles of Syndicate Members</u> You will find that there are a number of roles that have to be played by members of a syndicate which is to produce imaginative but thorough work: in particular there is a need for

- The 'ideas generator',
- The critic,
- The synthesiser, who tries to make sense of what is happening,
- The writer,
- The planner.

It is not necessary to have a 'leader' in the military sense, in fact if a dominant personality takes charge at the outset, the results of the syndicate are likely to be rapid, neat, but superficial and lacking in self-criticism. Each of the roles noted above needs to be present, and it may be that one person may play more than one, or that several people will contribute in the same role.

A valuable part of working in syndicate is experience of how the individuals in it can take up the role which have to be played, and indeed how a lack, or an excess, of individuals prepared to play one of them, can lead to difficulty. This is also one of the skills of a systems analyst - how to operate in teams (it is a skill which seems to be better appreciated in the United States than the United Kingdom). If the directing staff agree, an interesting appendix to the syndicate's submission on the case may well be an account of how it operated and developed its functions during the study.

The 'silent member' of a syndicate is not necessarily a dead-weight, if he participates: a nod, or a smile, or getting a calculator or cups of coffee at the right moment may be a powerful contribution, making the protagonists feel that their more extrovert activity is sup-

ported and approved by the group. However, the silent
member who turns up late, or not at all, or who radiates
boredom or disapproval, is really disruptive. The rest
of the group must find out what is wrong and either get
him to become a real member (even if in the role of voluble
critic) or else tell the directing staff about their problem.

Methodology

We now turn to the strategy the syndicate can adopt.

1 Firstly, read the case through, individually.

2 Talk about it in an unstructured way, and make
  notes on ideas and queries which arise. Do not
  be in a hurry to jump to a pat answer. During
  this stage you will be also learning about each
  other and your individual potential contributions.

3 Next attend the first tutorial session with the
  directing staff and note down answers to your
  questions and those of the other syndicates.

4 After this comes your first crucial syndicate
  meeting at which you work out your overall
  approach to the problem. Formally you must
  agree on

- What, as real people, are the members of the
  syndicates hoping to achieve by working on
  the case? For example

    a good assessment, leading to increased
    promotion chances or a good degree,

    a better understanding of the systems
    analyst's job etc.

- Who is the syndicate pretending to be? For
  example, a team of outside consultants.
  What is this 'simulated team' trying to
  achieve? The sale of their services?
  Increased reputation?

- To whom is the 'simulated team' addressing its
  report and presentation? For example, to a
  'simulated managing director' or a 'simulated
  stores clerk'. What are the real motivations
  (profit, getting home on time, etc.) of this
  individual or individuals?

  (It is important to distinguish between the sponsor -
  the person with money who sees the computer system as
  an investment to carry out better a function for
  which he is responsible - and the user - the person who

carries out the function day by day and is now perhaps going to use a computer. A report addressed to a sponsor will stress the benefits of the improved functions: it will be task-oriented. A report addressed to the user should convince hime that his job will be easier (or, indeed, more demanding, but interesting!) and should set his mind at rest if possible on such questions as redundancy, pay, organisation, and working hours and conditions. It will be people-oriented.)

What sort of report is the syndicate preparing? For example

- An initial study (sometimes called a feasibility study) which indicates the opportunities for improvement in the area under review, through the use of computer methods (and of others, if they occur), or

- a statement of requirements which is much more detailed, and describes exactly how a new computer-based system will operate, and how it will be implemented.

5   Next the syndicate should consider its own strategy and draw up a timetable working backwards from the day when the report/presentation is due.

Allow plenty of time for final writing up or preparation of presentation material.

Rough out the main sections to be covered: these will nearly always be

- Introduction including a re-statement of the team's terms of reference.

- Recommendations: short term, long term: computer-based, other.

- Present system and assessment of how well it meets its goals.

- Proposed new system: how it will be used, with several typical examples, and how it will meet the system goals better.

- Implementation plan.

- Economic evaluation.

- Technical section: equipment and programs needed.

6 Allocate the sections above to be written by individuals.

7 Still working backwards, set target dates for agreement on what will be the main proposals.

8 Arrange brainstorming sessions - see below.

9 Set up a project log which records discussions, notes, and queries.

Brainstorming

Now the syndicate has a plan of campaign and people know their individual areas of responsibility. But the solution must be a syndicate solution: each section of the report must hang together to make a coherent whole. So the next activity will be 'brainstorming' sessions: every member of the syndicate is free to suggest any idea that seems relevant, and the other members must, at this stage, build on such ideas, however outlandish, rather than demolish them.

As the sessions progress, the main proposals will begin to crystallise: a particular new system will emerge, with its characteristic advantages, but also implying certain equipment, changes in existing working arrangements, and implementation tasks, such as the specifying, writing, and testing of programs.

Evaluation of brainwaves

These activities take time and money, and so the overall value of introducing the proposed new system has to be worked out by the following steps

New system brainwave,

Check out benefits to sponsor (profit, etc.)

Check out that it can be operated by the users!

Work out technical details of system,

Determine what equipment will be needed, when,

Determine what tasks have to be done for the system to 'go live' e.g. program writing,

Prepare time-schedule,

Compute cash-flows and project value.

It is the art of the systems analyst which enables him to make good guesses at times and costs, and so give the sponsors a reasonable estimate of a project's value. Of course the actual costs of program development will not be known until later, but it is possible to make global (if crude) estimates on the basis of some rule of thumb like

>'This function will need four big programs, one hard and three medium.'

>'A big program means in our terms 4000 statements. We use a factor of 1.5 for hard ones, so the whole function needs 4.5 x 4000 = 18 000 statements.'

>'Systems effort is calculated as one day for every 10 statements - to include programming, documentation, meetings, etc..'

>'So the total effort is 1800 man days. At £50 a day thus is £90 000.'

Presentation

The syndicate now has some ideas which have stood the test of feasibility analysis and financial analysis. They should be ruthlessly whittled down to not more than three proposals which will appear in the final presentation as follows.

>'We recommend proposal A. Proposals B and C are alternatives which are feasible, but are not as good as A because ...' (This convinces the readers that the team has done its homework, and they may be glad of B or C if A turns out to be unattractive for reasons outside the team's terms of reference.)

or 'We recommend a staged approach. First obtain some early benefits by adopting proposal A. Then introduce the more ambitious proposal B. Or, if the more detailed cost/benefit analysis which will be a byproduct of A shows that C is preferable, introduce C instead of B.'

The syndicate can now split up into individuals writing their own section, with a certain amount of cross-checking: for example, the member writing the detailed implementation plan may discover that the early estimates were too optimistic and have to call a panic meeting of the group to see whether the revised times and costs still left A the best proposal.

All members should read the completed report and check for errors, inconsistencies, and poor style or spelling. If a verbal presentation is required, they should rehearse it with a few good visual aids to back it up.

Finally, submit the report and listen to how the other syndicates approached the problem.

4
Minims Restaurant

Minims Restaurant is owned by Universal Facilities Organisation (UFO), whose head office is in Phoenix, Arizona. UFO owns a variety of businesses world wide, including a pearl fishing co-operative in Japan and a distillery in Scotland. UFO's expertise lies in finance, and it has an IBM/370-158 in Phoenix which is fed with monthly operating data from each subsidiary, sent by small low-speed terminals at each location. (The operating data report shows planned versus actual this month and cumulative for the year: revenue, operating costs, capital expenditure, cash at bank.)

Minims has the following staff

	£ p.a.	
General Manager	7 000	
Assistant G.M.	5 500	
Chef	5 500	
Assistant Chef	4 000	
Cooks (4)	12 000	
Storeman	3 000	
Porter	2 500	
Head Waiter	3 500	
Barman and (9) Waiters	20 000	+ tips
Cleaners (3)	6 000	
Doorman	2 000	+ tips
Cashier	3 000	
Accountant	4 500	
	78 500	

Minims is an old fashioned building in St James's which costs £50 000 p.a. in rates, rents, lights, heating and maintenance. This figure includes £1000 on stationery, etc., for bills and stock records.

They serve 200 lunches a day during the week (average £2.50) and 300 dinners on Friday, Saturday and Sunday (£6.00).

$$200 \times 7 \times 52 \times £2.50 = £132\ 000$$
$$300 \times 3 \times 52 \times £6.00 = \underline{£280\ 800}$$
$$\text{Total } \underline{£462\ 800}$$

Approximately 20% of meals are paid for by credit cards which Minims have to pay a 5% discount on, to the credit card company. 30% are special company meals on contracts or banqueting arrangements, for which 25% discount is allowed.

$$\text{Total discounts: } £462\ 800 \times (20\% \times 5\% + 30\% \times 25\%)$$
$$= £462\ 800 \times 8\tfrac{1}{2}\%$$
$$= £\ 39\ 338$$

The lunches and dinners contribute revenue in the following way

	Contribution to revenue %	Contribution to costs of raw materials %
Meat } Fish }	30	20
Wines, etc.	35	50
Vegetables	15	15
Other	10	15
Service	10	–

It will be seen that raw materials requiring extra preparation are charged at a proportionately higher rate to the customer.

The overall cost of raw materials amounts to £250 000 p.a. The raw material cost per lunch or dinner actually sold is estimated at 35% of the selling price. Any discrepancy between this and the annual cost of raw materials is attributable to kitchen wastage

meals returned by customers and not charged (refusals)

failure to charge correct price for meal or waiter theft

loss of raw materials through going off, etc.

theft of raw materials

The overall financial picture of Minim's can thus be represented as follows

Revenue	£462 800	Raw materials and losses	£250 000
less discounts	£ 39 338	Staff	£ 78 500
		Overheads	£ 50 000
		Operating surplus	£ 44 962
	£423 462		£423 462

The above figures exclude excise and tax, which may be ignored.

The financial return is thus £44 962 on a net revenue of £423 462 or about 11%. This is regarded as too low by UFO, which wants to see at least 18%.

Current Restaurant Procedures

Use of tables: tables may be booked by anyone for any number of people. If the table is not claimed within 15 minutes of the agreed time, the head waiter may re-allocate it to anyone waiting. Bookings are taken by the cashier (where the telephone is) on loose sheets which are passed to the head waiter, who files them roughly in date/time of table use, on a clipboard. After a table is taken he throws away the sheet. No charge is made for booking, but the head waiter tends to get tips from people when they take up a reserved table (whether they booked it themselves or not).

The effect of the booking system is to reduce the utilisation of tables and waiters by about 20%, but the head waiter argues that many old customers would be lost if they had to take their chance. It is possible that a large proportion of these are 'special company discount' customers.

When the customers are seated and have made their choice from the menu, a waiter takes their order on an order form which he tears off a pad. He places this order form over another plain pad consisting of carbon (pink) carbon (green). This pad is called the Rainbow. He writes in the customer's orders and the prices. The order top copy is for the customer and will contain all his food and drink consumed: but the waiter completes each new course, or drinks order, on top of a new Rainbow, so that he has a pink/green pair for drinks, entrée, sweet-and-coffee. He takes the pink to the bar or kitchen: this authorises them to prepare the items ordered. He then collects and 'pays' for them with the green, picking up his pink again.

When the meal is over the waiter adds up the bill, adds 10% service charge, and gives the bill to the customer. The customer will

> pay the waiter and keep the bill
> or pay the cashier by credit card and keep the bill
> or sign the bill, showing a 'special discount customer' authority.

The waiter hands in his top copy and the pinks, and money/credit card if any, to the cashier, who checks the total, and whether it is roughly in line with what the customer had (by reference to the pinks). Cash and credit cards are dealt with through the till in the usual way; signed bills are collected for billing the 'special discount customer' later.

If a customer refuses to pay for some food, the assistant manager initials the bill. The manager himself counts the money/credit card slips and reconciles the totals after closing time, and puts the money in a safe ready for passing to the bank the next day.

Kitchen and Bar Procedures

On receipt of the pink, the assistant chef looks to see whether there is going to be enough of each item already in course of preparation. If so, he takes no action, but if there is an item that needs to be started from scratch he will initiate it by telling the cooks, always rounding up the quantity to an amount worth cooking, except near the end of lunch or dinner when he will try to avoid unsold food. This process is done by the assitant chef's 'feel' for the situation. Certain items may be withdrawn from the menu or others may be in over-supply - this information is passed to waiters to encourage customers to order in a helpful way ('just a word in your ear sir - there's a lovely piece of Scotch salmon we are keeping for our old customers') - (Yes: the tin has been open three days now and we must get rid of it!)

When the food is ready he will instruct cooks to make up the orders on plates and will show the 'pinks' in a small window to tell the waiter that the order can be collected.

He then exchanges the pink for green and gives the food to the waiter.

At the end of the day (or next morning) the chef himself goes through the greens and totals off the number of each portion of each kind which was served, for example 87 rump steaks. He then does a subjective check against the issues made out of the storeroom to the kitchen, recorded on the Grub Chitty (see below). Thus if 150 raw rump steaks were actually issued, he will try to find out why only 87 reached the customers, or why a hundredweight of potatoes were issued, when every customer only had salad.

Once a week he goes into the stores and does a check on stock levels, and arranges for a re-order if any raw material is running below target levels. There is also a fast-move section (cream, fresh fish, vegetables, etc.) which is turned over daily. Orders are raised by the storeman and counter-signed by the chef. A rather similar scheme operates for the bar.

The cooks make out Grub Chitties whenever they need more raw materials to meet the cooking schedule dictated by the assistant chef as above. They pass these to the storeman in return for the raw materials.

The back door of the storeroom is kept locked except when deliveries are made, when the assistant chef or chef opens it and inspects the goods and signs the delivery acceptance note. Another copy is signed by them and passed to the assistant manager who keeps them until suppliers' bills come in, when he pays them.

YOUR BRIEF

UFO has sent a team of systems experts to examine Minims with the following brief

> To what extent might fraud and ineffective controls be causing excessive losses, and if the figures given hide the worst possible situation, what operating surplus could be achieved?
>
> To comment on current procedures
>
> To propose new procedures that will in the first instance pinpoint the cause of losses, and secondly reduce them.

You are invited to play the part of this team.

Possible Equipment

UFO's Electronautics Division markets an ingenious device intended for Point of Sale applications. It consists of a

CPU with up to 10 terminals. The CPU has access to a floppy disc holding 2m bytes, and has adequate memory for handling stock control, billing, etc. The disc and/or CPU fail once a year.

Each terminal consists of a VDU, a normal typewriter keyboard with a separate adding-machine-like numeric keyboard, a badge reader, and a printer which can produce multi-copy forms and a tally-roll. Further options include an OCR reader for forms printed on the same type of terminal with digits in defined boxes, or for marks written by hand in boxes. The OCR reader rejects 0.1% of the marks and mistakes 0.01% of them.

Another feature is the UFO Magic Star Identifier: rolls of backing paper can be bought on which are removable 'stars' about the size of a 50p piece. Each one has a number printed on it, ranging from 1 to 99999. The terminal can be equipped with a Magic Star Wand which consists of a sensor on a ten foot cable, which, when placed on or near the Magic Star, communicates the number to the terminal and so to the CPU. The Stars have a parity check built in but there is a 2% chance of an (announced) error and a 0.1% chance of an unannounced error. Stars enable objects to be identified quickly and automatically.

Terminals can be arranged to be usable only by people who have the right badge, or the badge can be used just to identify the person concerned. They can also have keys which lock out unauthorised personnel, or unauthorised transactions.

UFO will make this equipment available to Minims at a specially reduced rate.

CPU, software	£10 000	
2m byte disc	1500	
2m byte backup cassette	1000	if required
Terminal VDU, keyboard	750	each
Badge reader	150	
OCR reader	500	
Printer	500	
Magic Star Identifier	500	(stars cost £1 per 1000)

Moreover UFO will allow Minims to pay only one-quarter of these figures a year for four years if they so desire. But Minims must identify and if possible reduce the losses which are occurring.

If the equipment breaks down, some fall-back system must be

arranged. There is another one, but in Brussels, so it would take a week to use. It may be worth providing duplicated systems to some extent if the costs of failure are likely to be high, and alternative manual fall-backs are too expensive or awkward.

A further point should be considered. UFO will be installing a teletype and modem at Minims so that they can communicate financial facts and figures to Phoenix. No charge will be made to Minims for this, and Minims can use the terminal for other purposes, within reason, for example, analysing figures. (Use would be limited to half an hour a day.) Other equipment, service bureaux, etc., may be used: details on request.

5
The Dental Supply Company

HISTORY OF EVENTS

Universal Facilities Organisation (UFO) is an American conglomerate with interests in a wide range of industrial and commercial areas throughout the world. Its main interest is in small to medium-size operations. Unlike many American corporations, UFO has little interest in anything but financial control over its subsidiaries.

Ten years ago, UFO bought the Dental Supply Company, a British firm established by the son of a dentist in 1921 to manufacture dental equipment for the UK market. The company now has about 20 per cent of the UK dental 'hardware' market and has added the sale of a range of bought-in dental consumables since being acquired by UFO.

September 1980

Universal Facilities Organisation, which owns Dental Supply (DSC), decides that DSC should expand its activities by acquisition. On retirement of the DSC chairman, the current MD becomes chairman and a marketing specialist from UFO's Phoenix, Arizona HQ, is appointed MD: he is Tom White, age 34, graduate of the Harvard Business School. Tom's terms of reference are given in Exhibit A. A description of DSC's present activities is given in Exhibit B.

November 1980

Tom White identifies Orthodontic Consumables Co. (OCC) as a possible acquisition. He calls a team from UFO HQ to evaluate the financial benefits of DSC taking over OCC, and operating as an integrated new concern. Their report is given as Exhibit C. Tom White recognises that a key factor in making a success of the acquisition will be the rationalising of OCC's order-taking systems with those of DSC. In particular, a physical move of all OCC staff from Hammersmith to DSC's location in Slough, and a thoroughgoing reorganisation of the ordering system at DSC to include OCC orders, probably using computer systems.

December 1980

Tom White calls in Mr. Johnson, manager of DSC's DP department, and finds him unable to suggest ideas for using computers. Tom turns to a software house, Rationalistics, which is part of a larger consultancy, Systems Consultancy Ltd (SCL).

31 January 1981

Rationalistics propose an online system with a minicomputer, visual display units (VDUs) and their own software. Their report appears as Exhibit D. They estimate they can get the system up and running in three months.

10 February 1981

Having thus checked the feasibility of a new ordering and DP system for the proposed new organisation - DSC integrated with OCC - Tom goes ahead and buys OCC from its owner, Sam Jones. An announcement is made to the staff of each company asking them to work as before pending a reorganisation. DSC holds a senior staff meeting (Exhibit E).

17 February 1981

Tom White receives two worrying letters. (Exhibits F and G.)

> The staff of OCC, who are not unionised, express their anxiety about the reorganisation, since they have heard rumours that they will be asked to move to Slough. They ask for assurances that they will not lose their jobs, and they demand substantial compensation (£5000 is mentioned), if they are asked to re-locate.

> The local union convenor in Slough for the union representing the DSC order clerks, writes to say that he has been informed that there is a proposal that his members will be expected to use a computer and wear telephone head-sets under the new order-taking system, and that their jobs may also be in danger from automation and from the arrival of non-unionised staff from OCC. His members want assurances that these things will not happen.

Tom is very concerned: the whole basis on which the DSC-OCC merger was to be profitable is being undermined.

However, when he originally spoke to SCL, the senior consultant had told him that Rationalistics, the software division, was only one of several consulting divisions, and so he now commissions SCL to advise him. A small team of top-level experts arrive and are given copies of all the papers in the case, and set to work to advise Tom on what to do next. Their terms of reference are set out in Exhibit H.

EXHIBIT A

TOM WHITE'S TERMS OF REFERENCE AS MANAGING DIRECTOR OF DENTAL SUPPLY COMPANY

Dear Tom,

As we discussed, I have pleasure in confirming that the Board of DSC, on UFO's recommendation as majority shareholder, has unanimously agreed you should be Managing Director in succession to James Smith, who becomes Chairman, following the former Chairman's retirement. Your appointment is from 1st October 1980 and is officially permanent although it is my hope that after about three years, when you have put DSC soundly on a new growth path, you will take up further senior appointments in UFO.

Your brief, which is agreed with the DSC Board, is:

> To direct everyday operations of DSC through your Finance, Marketing, Administration, and Production executives.

> To enlarge the worth and revenues of DSC by increasing sales to dentists both of existing DSC products, and of other companies' products, through acquisition.

> To secure cost reductions, through economies of scale in marketing and administration, by rationalising and centralising these functions after acquisitions.

Board Structure. As agreed, a new post of Marketing Director is created, and will be filled by Henry Black, your present deputy in Product Marketing, UFO. Sales and order processing will report to him (previously the sales manager reported direct to the MD, and the Order Office to the Administration Director). The Board is thus

> Chairman: James Smith

> Managing Director: yourself (UFO nominee)

> Marketing Director: Henry Black (UFO nominee)

> Finance Director: William Jones

> Production Director: Thomas Edwards

> Administration and Company Secretary: Herbert Williams

Non-executive Director: Cyrus P. Gerhardt (UFO)

James Smith will not oppose you, I am sure, unless he genuinely feels that your proposals are unworkable. In such a case, and assuming the other local directors all oppose you, the issue will be referred to the UFO main Board: I hope, though, that this will not happen.

>Very best wishes for your success,

>Julius L. Malone   President.

EXHIBIT B

THE DENTAL SUPPLY COMPANY

Dental Supply Company (DSC) is an old established firm in Slough, which sells a range of goods to hospitals and dentists in seven main product groups:

1. Dentists' chairs, three models. Accessories such as head-rests for the chairs.

2. Drills, three models (high speed,...).

3. Pedestals and washbasins (two models each).

4. Cabinets, worktops, mortars, patient record files, etc. Ten products.

5. Waiting room furniture. (Bought outside).

6. X-ray apparatus (electrics and camera bought in, mounting, pedestal made by DSC).

7. Drill bits, X-ray film packs, patient record cards and stationery, and other low value, high usage items: 20 products.

Groups 1-6 are fairly expensive (average value £625) and are ordered fairly infrequently. Typically a DSC customer orders an item from each of these categories about once every 1.5 years, but he is likely to order a group 7 item once a month.

DSC has 8000 customers. This is static, with new customers (500 a year) being offset by an equal number of losses due, it is believed, to dissatisfaction with DSC's slow response to orders in all groups.

Revenue can be analysed as follows:

Groups 1-6	8000 x £625 x 6/1.5	=	£20 000 000
Group 7	8000 x £50 x 12	=	£ 4 800 000
			£24 800 000

Orders are handled by mail. Salesmen visit customers to tell them about DSC products, but do not take orders. Advertisements appear in The Dentist and similar journals, and mail shots are sent out.

At Slough, incoming orders are sorted by group, and there is a clerk for each group 1-6, and four clerks for group 7. If a customer orders items from more than one group, his order is copied and each clerk involved is given a copy, and deals with it independently - to the point that an order for several different products will be shipped in several consignments.

The average number of orders a day is

Groups 1-6   8000 x 1/(230 days x 1.5) x 6 = 139 approx.

Group 7   8000 x 12/230 x 1   = 417 approx.

Accordingly, each clerk in groups 1-6 deals with around 23 orders a day, although each order takes several weeks to process in its entirety:

> Credit check on customer. Very few dentists or hospitals turn out to be bad debtors, but it is estimated that £1000 a year would be lost if this check were omitted. It involves sending a slip to the Accounts department, to be returned after about four days with an indication of the customer's creditworthiness, based on the state of his account. New customers are checked by sending the slip to a credit agency which takes five days.
>
> The product ordered by the customer is checked with the catalogue and the right code found. It may be necessary to ring or write to the customer to ensure the correct product has been identified. This introduces an average delay of two days.
>
> A check has to be made on the delivery position of the goods ordered: this involves sending an enquiry slip to Stock Control, where the delivery position is examined and if satisfactory, the goods are allocated, pending the final instruction to ship. This procedure takes four days.
>
> When all the information has been obtained, the clerk drafts an acknowledgement letter quoting the exact product specification, price, delivery, etc. One copy goes to the customer, one to Stock Control, one to Accounts and one to DP.

Group 7 items are handled differently. A group of four clerks handles over 100 orders a day each, using a simpler procedure.

> A card is pulled from a customer-card file arranged in alphabetical sequence. If no card is present, Accounts are notified and they authorise DP to pre-

pare one. Sometimes the reason why the card is missing is simply that it has already been pulled: this creates some confusion and delay.

One or more product-quantity cards are pulled from a tray, each representing, for example, 144 patient record cards type A6.

The card-set (customer card followed by product card(s)) goes to the DP department for them to prepare order documents.

When DP have finished with the cards, after about five days, the group 7 clerks collect them and refile them.

The DP department does not handle group 1-6 orders at all until after the goods have been despatched, when Accounts prepare an invoice, a copy of which goes to DP, and triggers the punching of an invoice card which is added to a card file. This card file is essentially an open-item sales ledger, and is tabulated monthly to produce statements. It is also sorted and tabulated in various ways to give sales statistics, on average, a month after despatch of goods.

DP does however produce the invoices and invoice cards for group 7 items in one operation, from the card sets sent by the Order Office, as described above. The computer, a simple card machine, adds up the quantities and prepunched prices, card by card, and punches a summary card for each set.

Accounts may send credit notes, which are sorted into the invoice card file. Cash receipt cards are similarly sorted into the file, and every month the sales ledger listing is marked off by Accounts to show cash, invoice, and credit cards to be removed. These should of course balance to zero and this is checked with a special computer run on the removed cards.

It is noted that customers are dissatisfied with the way that they can get billed for group 7 items before the goods arrive, and this bad feeling contributes to the numbers of customers who leave DSC each year.

Organisation. The ten order clerks and five typists report to a supervisor who used to work for the Administration Director, but who will now report to the new Marketing Director. DP consists of a manager, two analysts, two programmers, six machine operators, and eight punch operators. DP comes under Administration.

Accounts come under the Finance Director, and the Sales Accounts section has 25 clerks and a supervisor. The Sales department reports to the Marketing Director, and has a manager, a sales research officer, an advertising manager and 20 salesmen who pay calls on customers on average once every two years or so, but do not take orders.

The Production Director runs Manufacturing and Purchasing, the details of which need not concern us, except that he estimates that if he had better information about demand by product, he could cut his costs by up to 5 per cent.

DSC's overall financial situation may be seen from the following figures.

DSC's Operations	£
Manufacturing and purchasing costs	20 000 000
Premises, plant, other fixed costs	1 000 000
Directors' and senior executives' costs	250 000
Order Office (16 @ £5000)	80 000
Sales Office (23 @ £7000)	161 000
DP (19 @ £5000)	95 000
Sales Accounts (26 @ £5000)	130 000
Other departments	200 000
DP rentals	20 000
Bank charges	864 000
Operating surplus	2 000 000
Revenue	£ 24 800 000

Note: Delivery costs are included in Manufacturing, etc. Bank charges are interest on overdraft, at 10 per cent.

EXHIBIT C

SUMMARY OF FINANCIAL ADVANTAGES THAT WOULD FOLLOW FROM THE ACQUISITION BY DSC OF THE ORTHODONTIC CONSUMABLES COMPANY (OCC)

CONFIDENTIAL

To   Tom White, Managing Director, DSC
From   David Wiener, UFO Financial Analysis Unit
Date   15 November 1980

1. Recommendation. We have made a confidential investigation into OCC as you asked (2 November 1980). Our conclusion is that revenues could be increased and cost reduced by your acquiring OCC and moreover, by selling OCC's Hammersmith office and factory, you will obtain £1 500 000 towards the costs of acquisition. Details follow below.

2. Proviso. To make a success of this acquisition, you will need to move rapidly to integrate procedures between DSC and OCC in the fields of order-taking, marketing, and DP. While this is happening, morale at both companies must be maintained to avoid losses.

3. Note. Although this was strictly outside my terms of reference from you, I observed that procedures at DSC are, even without the possibility of a merger with another company, in urgent need of revision, and benefits could be obtained from improvements. However, an even better result will be obtained by changes which cover both these improvements and the measures that arise from the proposed integration.

4. Description of OCC. OCC is a 'one-man firm', started by Sam Jones, who is 100 per cent owner and MD/Chairman, on money he obtained from scrap metal dealing after the Second World War. The OCC premises are a warehouse he bought cheap when cracks were found in the walls, and he has converted it into a factory and offices producing and selling

        Mouthwash
        Dental amalgam
        Disposable swabs and spatulas, etc.
        Disinfectant
        Other consumable stores used by dentists.

Sam Jones foresaw the increase in dental care that resulted from the introduction of the Welfare State. The production processes are very simple and are carried out by unskilled workers, largely female immigrants. His ordering system is particularly crude and effective.

He employs four order clerks, who wear telephone head-sets, and type customers' orders on to a multi-part carbon set at their dictation. The set contains a copy for despatch, one for the customer, which serves as advice note and invoice, and a third copy. Directly below the order office is the store room, containing open bins of each product ready for sale. The order clerks call down a speaking tube to stores clerks who check from the bin whether the order can be met. If it cannot, the clerks send down the order form set and it is put at the bottom of a pile of such forms by the bin; when stock comes in, a stores clerk sends back up to the order office as many of the orders in the pile as can be satisfied, and the order clerks ring up the customers and go on with the transaction: they price each order and tell the customer the value, delivery date and method (post or van). The order set now goes down to the stores clerks who organise packing and despatch. A copy of the order set is put on a spike in Accounts, representing an invoice awaiting payment. There is a row of spikes, each representing a day's invoices: Accounts remove them as they are paid by customers, and tell the order clerks of the 'bad' customers who have unpaid invoices on 'old' spikes, so that they can refuse further orders until the old ones are paid. However, the clerks have discretion to accept orders if they think there is a good reason why the customer's old invoice has not yet been paid.

OCC receive about 800 orders a day, split between the four clerks. Their average value is £25, so the annual revenue is about

    800 orders x £25 x 230 working days = £4 600 00

OCC has the same sized market as DSC - 8000 customers - and so they are placing orders on OCC every 10 working days.

Sam Jones works on large profit margins. He has no sales force, but is a well-known personality who uses unconventional and distinctive promotional methods, like his habit of inserting vouchers at random in customer consignments, which entitle the recipient to free holidays, or dinner for two at the Hilton: he calls these 'Goodies for the Snapper people' and although the dentists affect to despise this approach, he sells a lot of mouthwash.

Much of the above information was obtained confidentially

by informal contacts I do not wish to reveal.

Sam Jones's trading results are hard to obtain in any detail since he publishes only the minimum required for a privately owned business. However, his operating results can be summarised as follows:

	£
Manufacturing and purchase costs	2 500 000
Overheads	500 000
Operating Surplus	1 600 000
Total revenue	£ 4 600 000

5. Analysis of acquisition project. Sam Jones is a very rich man but is concerned about the future of OCC after his retirement or death. He would certainly accept a deal in which OCC became part of DSC, if he received cash and/or UFO shares, paid into an off-shore account where it would attract little UK tax, to a total value of £2 500 000.

DSC could then sell OCC's Hammersmith site for £1 500 000 and move OCC's activities to Slough, where the order-taking, accounts, and DP systems would be integrated, and OCC's products held in the DSC warehouse.

Even if all OCC staff had to be retained and no benefits were realised from consolidated and improved order-taking and marketing procedures, the OCC acquisition would cost a net £1 000 000 and gain a business whose annual operating surplus, £1 600 000 would pay for the acquisition in a single year.

However, the situation should be even better, in that DSC staff will absorb the OCC clerical and administrative functions, so saving the bulk of OCC's £500 000 overheads. This assumes that DSC's Order Office, Sales Office, Accounts and DP can improve their procedures to take on the extra OCC load without an increase in staff, which one would assume feasible, since OCC has shown that four order clerks can handle 800 orders a day between them even without computer assistance.

Apart from the economies to be obtained from administrative rationalisation of DSC/OCC, we should mention that the combined company would be able to present a product line which would give our customers a complete range of dentists' requisites, to be delivered by a unified distribution system.

6. Conclusion. We believe you should move quickly before Sam Jones realises the value of his Hammersmith site, and considers other offers.

EXHIBIT D

RATIONALISTICS' REPORT

Dear Mr White,

Thank you for your invitation for us to propose a computer-based system to support your centralisation of DSC/OCC order-taking and sales accounting functions.

We recommend you commission us to procure and install an ICM 4G computer, procure or develop software for your system, and implement the system described below.

We can do this within three months of receipt of order. The benefits you will obtain will exceed £200 000 per annum - see the section on economies below. The costs are £127 500 one-time, plus recurring charges for maintenance of equipment, of £7800 p.a. We assume that other costs such as stationery will remain much as at present.

Outline of proposed new system.

We now describe how the new system will operate. Orders for DSC/OCC products will arrive by mail or telephone, and will be routed by the supervisor to the first free clerk of eight. Each clerk will thus be trained and equipped to deal with orders for either company, any product group, and by mail or phone. Each will have a telephone head-set and a VDU and keyboard, connected on-line to the computer.

In the case of telephone orders, the clerk asks for the customer's own telephone number and enters this number through the keyboard. The computer will search for the customer who owns this number, and will display his record on the clerk's VDU, as a check on the name and address. The record also shows whether the customer's credit rating is satisfactory. In those cases where a telephone number is not available, the clerk can also retrieve the customer record by keying in the first three characters of his name, ignoring vowels, and the computer will then respond with a VDU display of all customers whose names could match. In the case of common names like JNS for Jones, Jonson, etc., we expect that the list of possible customers will not exceed the capacity of two VDU pages, and in the vast majority of cases the required customer name will be on the first page. The clerk then types in the line number of the customer concerned.

Mail orders are dealt with in much the same way except that the clerk works to the customer's letter rather than the customer's voice over the head-set.

After the customer has been identified and his credit-worthiness established, the clerk selects a list of the main DSC/OCC product groups on the VDU and keys in the line number of the group in which the item ordered belongs, and this causes the computer to display on the VDU a list of all items for this group, from which the clerk makes a selection. This in turn causes the item's own record to appear, showing price, delivery, etc., as well as a full description for checking with the customer (or his letter).

The clerk enters the quantity required, and can then go on to another order, or other items on the same order. After an order has been produced, a complete despatch set is printed on the line printer: stores copy, despatch copy, customer copy, receipt note, and, for pre-invoiced goods, an invoice copy.

If there is no record for the customer on file, the clerks refer the order to the sales accounts section who raise one, when satisfied with his references: the order is then dealt with the following day, although in special cases and with the authority of the accounts manager, the customer account can be set up there and then through the keyboard.

We have described, in outline, the order office procedures. In the sales accounts section, there will be two VDUs which will be used for cash posting: as cash is received, the clerks can look up the customer's account and determine which invoices can be cleared by the payment, and may allow the customer to regain creditworthiness if previously 'on stop' for slow payment. The VDUs can also be used for entering agreed credits.

System Benefits

We now describe the benefits of this system. Firstly, there is a considerable saving in staff. There are now four OCC clerks, 15 DSC clerks, including typists, and two supervisors. Under the new system there will be one supervisor and eight clerks, and no typists, so that 12 staff are saved. If these are charged at £5000 p.a., the annual saving is £60 000.

In addition, we believe that there will be a significant reduction in the number of customers leaving DSC through dissatisfaction with slow ordering procedures. If this figure is halved, we stand to gain the revenues of 250 additional customers each year, hence revenue increases by 250/8000 x £24.8m = £775 000 p.a., and the contribution to overheads involved is £775 000 x (£24.8m - £20m)/£24.8m = £150 000 p.a.

In sales accounts, procedures will be improved and manual

typing reduced, so that we estimate that at least two out of 25 staff will be saved, giving an annual saving of £10 000.

Hence we estimate the total financial benefits at £60 000 + £150 000 + £10 000 = £220 000 p.a.

System Costs

We have to set off the costs of the system against these benefits. We estimate the once-off costs at £97 500 for the computer and air conditioning, etc., £20 000 for Rationalistics software, and £10 000 for training and implementation costs. This amounts to an investment of £127 500, much of which will be eligible for investment grants and allowances.

Choice of Computer Equipment

We now turn to the exact configuration of computer you need. We recommend the ICM 4G, with a central processor with 256k and a 0.25 microsecond cycle time. This is needed to hold a real-time operating system and applications programs, buffers, etc., for the 10 VDUs (8 in the order office, 2 in sales accounting).

This unit costs £20 000. We also need 10 VDUs at £1000 each, two disc drives to hold between them 8000 customer records and product data, etc., two printers, one card reader, and multiplex and selector channels. We also recommend the purchase of 25 disc packs. To this total we add an estimate of 25 per cent extra for air conditioning, etc. Maintenance is based on 10 per cent of the overall purchase price per annum.

Schedule:	£
CPU ICM 4G 256K, 0.25 usec	20 000
10 VDUs	10 000
2 disc drives	20 000
2 line printers	14 000
card reader	5 000
channels	4 000
25 disc-packs	5 000
Sub total	78 000
Air conditioning, etc. @ 25%	19 000
Rationalistics fees	20 000
Training, implementation	10 000
Total one-time costs	127 000
Recurring costs (maintenance) at 10% of sub total above	7 800

Note that stationery, floor space, etc., are assumed to remain essentially at the same level as before, except that the OCC order office will be released giving an opportunity for DSC to sell the premises.

Security Provisions

We should include in this report our recommendations on the security arrangements which should be built into the system, since you need a reliable system if it is to maintain and improve DSC's reputation. If individual VDUs, line printers, or discs drives fail, the configuration is such that the service can continue. However, if the processor fails, we have a fall-back procedure: the clerks enter details of telephone orders on note-pads, and when the system is restored, these, and mail orders, can be entered retrospectively. Since the system is in use only 8 hours a day for order-taking, there are 16 hours available for recovery in this way. If there is a prolonged computer failure, the discs can be taken to the service bureau at Watford with a 4G, and VDUs at your office connected to it by means of modems and dial-up lines. The speed of the VDUs will be reduced by remote connection, but not excessively.

Workload Calculation

We estimate that the system will cope easily with the workload, by the following calculation. The total number of orders a day is, on average, 139 DSC group 1-6, + 417 DSC group 7, + 800 approx. OCC orders, = 1356 approx. per day.

To process each order, the clerk will need to key in a telephone number, a product group, an item number, and a quantity, together with a few control symbols: say 20 key depressions taking not more than 10 seconds with practice. The computer responses will be not longer than 1 second each for four VDU displays, on average, and the rest of the transaction will consist of telephone conversation or letter handling, say 45 seconds, giving about a minute per transaction.

Eight clerks can handle 60 x 8 x 8 orders in an eight-hour day, therefore their total capacity is 3840 orders per day - comfortably more than the daily average of 1356. Indeed, it may be possible to reduce the clerical staff levels further.

When telephone orders start to peak at the end of the morning, mail orders will be ignored until the telephone load eases. The concept of clerk interchangeability makes this possible. It follows similar practices in GPO counter services, where economies were obtained by despecialising clerks.

Conclusion
==========

This letter summarises our detailed studies which are attached as appendices. We look forward to hearing from you, and hope that you will ask us to go ahead and install the system. Rationalistics have successfully commissioned 20 systems of this type to date, and our experience is your guarantee of a reliable, cost-effective, and punctual installation.

Yours sincerely,

J. T. Smith   Senior Consultant, Rationalistics Division
               Systems Consultancy Limited, London
               31 January 1981

EXHIBIT E

NOTES ON A MEETING HELD ON 10th FEBRUARY

Mr. White welcomed directors and senior staff of DSC. He announced the purchase of OCC, and explained that some secrecy had been necessary to avoid encouraging a competitive bid. However, the DSC Board had approved the purchase. Mr. White then explained that the full benefits of the acquisition would only be secured by a major reorganisation in staff duties and methods: in particular, the use of computers. Copies of a proposal from Rationalistics Division of SCL were distributed. He hoped that staff would welcome this opportunity to increase the scope of DSC, which would result in greater responsibility and rewards for those involved.

He explained the new procedures to be introduced in the Order Office and Sales Accounting, and the organisational changes that went with them.

The Chairman, James Smith, said that everyone would be excited by the news. He hoped that DSC would gain financially, but not of course at the cost of destroying the spirit of companionship that had been a characteristic of the DSC team, or of losing the loyalty of long serving employees. However he was sure Mr. White was aware of his concern.

The Finance Director, William Jones, said he was attracted by the very great improvements in revenue, profits, and assets that the venture offered. He was however anxious that any changes in the short term would not create dislocations in the business, so prejudicing the future benefits which he was sure were there.

The Administration Director, Herbert Williams, said he hoped that the clerks in the Order Office, now to work for Henry Black, the Marketing Director, would be as happy there as they had been with him.

The supervisor of the Order Office said that they were sorry to have to leave Mr. Williams, and needed time to think about the implications of the new system.

The Sales Accounts supervisor asked for time to read the consultants' report, and expressed his concern about the problem of credit control.

The new Marketing Director, Henry Black, said how pleased he was to be in DSC, and that he hoped that the Order Office would gain a new sense of importance from feeling themselves part of the selling team rather than 'backroom boys and gals'.

Tom Edwards, the Production Director, said he hoped the new system would give him more advance warning, rather than less, in planning manufacturing programmes.

Mr. Johnson, the DP manager, said that he was disgusted with the high-handed way in which an outside organisation's ideas were being foisted on the company. But for the fact that he had five years to go before his pension, he would resign there and then.

Cyrus P. Gerhardt, non-executive director, said he hoped that everyone would back up Tom White to make a real success of the venture.

Also present at the meeting:

Mrs. June Brown, Personnel Officer.

EXHIBIT F

LETTER FROM OCC ORDER OFFICE TO TOM WHITE

Dear Mr. White,

We are writing to you to make sure we get a fair deal. There are a lot of rumours flying round the place that you plan to move us to Slough, and most of us have husbands and families in Hammersmith. If you do try and move us, we shall want at least £5000 a head and even then we're not keen.

We expect you'd like to make us all redundant instead – but you had better know that we are a darn sight better order clerks than the DSC crowd.

But if you are thinking of sacking us, think again. We want a letter from you to say our jobs are safe. And another thing, many of us are coloured, and if you're not careful we'll have the Race Relations Board on to you.

Signed by all the clerks at OCC order office.

15th February 1981.

EXHIBIT G

LETTER FROM LOCAL UNION CONVENOR, SLOUGH

Dear Mr. White,

This is to give you notice that the appointed shop steward of your order office at DSC has laid before me certain facts which lead me to protest on our members' behalf, and to request an urgent meeting with you.

It has come to our attention that members will be asked to wear telephone head-sets, and to operate a computer. Our members have also heard that non-union staff are likely to join them at Slough, and that redundancy among DSC staff may occur. We are unable to accept any arrangement that will lead to redundancy and, furthermore, the wearing of head-sets is unacceptable - it is demeaning and inconsistent with the status of clerical staff such as our members.

I will telephone you to arrange a meeting with you to give us assurances which we can convey to our members.

Yours sincerely,

CWU Convenor, Slough

EXHIBIT H

TOM WHITE'S LETTER TO SCL

CONFIDENTIAL

Managing Director,
SCL

Dear

I need your help. As you recall, DSC has just acquired OCC, and I am going to make money out of doing this. Your Rationalistics Division proposed (31 January 1981) an on-line order entry system which I liked, and used as the basis for the organisational changes and new procedures I announced on 17th February.

It now appears that there is resistance to these changes among my colleagues, senior staff, and clerical staff at DSC and OCC. If I go ahead with your Rationalistics proposal I expect DSC and OCC to come to a standstill, and for my board to oppose me. If I do not move, the acquisition will have been a failure.

I want your best people to study the situation, and advise me what actions I should take. I am not particularly worried about fees, but let me know if your report will cost more than £5000. I want it soon.

Yours sincerely,

Tom White
Managing Director, DSC.

BRIEFS

Three different briefs can be given to the syndicates:

1. You are the original Rationalistics team: propose a system as it did in Exhibit D, but take account of the likely industrial/organisational effects so that your system is more acceptable, or less vulnerable.

2. You are Mr. Johnson, who has decided after all to suggest a development of his own existing simple computer system, so as to preserve his authority. Design the extensions necessary, trying to avoid real-time and VDUs, and draft a letter to Tom White which proposes them and climbs down a little from his attitude at the 10th February meeting.

3. Fulfil the brief given to SCL by Tom White in Exhibit H.

6
Bluebird Aircraft

Bluebird Aircraft manufactures a small range of light aircraft for the private pilot and clubs

    'Trainer' B100 2-place, 100 h.p. Price £12 000 excl VAT
    'Tourer' B160 4-place, 160 h.p.   £16 000
    'Super-T' B200 4/5-place, 200 h.p.   £18 000

The Bluebird range is designed to make maximum use of components which are already being mass-produced, particularly for cars. They use Porsche air-cooled engines and Fiat alloy door fittings, for example. Many other components are bought in, such as propellers, instruments, seats, etc., from other light aircraft companies.

To avoid being held to ransom by one supplier, Bluebird has, if possible, more than one source for every component. This creates a problem because the alternatives are sometimes not quite the same. For example, the Fiat door latches are fastened to the door jamb with three bolts, but the Automotive Products alternative part is fastened with four. The door jamb is a fibreglass component made by Bluebird itself, so that it has two varients, one drilled and reinforced for Fiat, one for AP door latches.

In some cases Bluebird has designed 'adapters' which enable alternative components to be fitted easily; for example, to fit a square-cross-sectioned alternative source (AS) instrument in the panel instead of the primary-source (PS) one, they stock or can make themselves a simple aluminium pressing on the lines shown in the figure opposite.

While this approach gives Bluebird the opportunity to obtain components at a reasonable price without the overheads of designing and making them itself, it creates other problems; in particular, the number of <u>combinations</u> of different components that could be used to make a Bluebird aircraft is very large. This means that Bluebird has to be good about specification control.

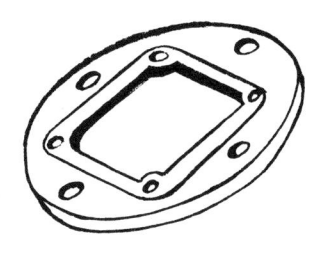

In addition, the certificate of airworthiness for each type includes limitations on which parts can be physically fitted in the same airframe, for example, although several sizes of commercially available tyres can be used, Bluebird must not use tyres of different sizes on left and right main wheels of the same aeroplane.

Further, some AS components weigh more than their PS equivalents, or less, and the adapters have to be taken account of, if used. The whole aircraft must weigh less than a certain maximum, and its centre of gravity must lie within certain limits (measured as distance from the rear of the firewall) which vary with all-up weight. Hence, although steel bulkheads or steel spars may be allowed, both together could be illegal.

Current Documents

Structure File   The specification for each aircraft type consists of a binder with loose-leaf A4 pages, divided into two sections: Structure and Limitations. In the Structure section, each component is broken down into lower-level components. Thus the aircraft as a whole (say B100) may be split first by a page S100

       S100.1   Airframe   1 off

       S100.2   Engine assembly   1 off

       S100.3   Hydraulics assembly   1 off

       S100.4   Electrics assembly 1 off

       S100.5   Electronics assembly   1 off

The page S100.1 will show the structure of the airframe, and so on.

At the lowest level, a component will be bought in complete, or manufactured by Bluebird. In this case the item's weight, distance from datum (firewall) and total moment are shown, together with the supplier's part number. If there is an AS this is shown too. Thus for hydraulics page S100.3 might show

S100.3.1   Hydraulic pump 1 off, 30lb x 18" = +540   Lucas
                                                                                                                                                L1234567
                                                                                                                                                PS

                                                               35lb x 18" = +630   Rotol
                                                                                                                          PQRST12345
                                                                                                                          AS

S100.3.2   Brake actuators 2 off, 40lb x −40" = −1600   AP
                                                                                                                          AP987654 X

S100.3.3   Brake assembly (controls) 1 off, ...

      etc.

Note that the third item will give rise to S100.3.3.1, etc.

It may be that the choice of an AS (there may be several, for example, AS 1, AS 2, ...) will force the choice of another component, for example, 'not (B100.1.2.3 AS 2 and B100.1.3.3 AS 1)' etc., as when two different components cannot both be fitted in a limited panel, for reasons of space. In this case, the note will be appended to the pages where any of the components in the rule are mentioned. Note that assemblies themselves may be PS or AS.

The Limitations section itself deals only with weights and moments, and is filed at the back. For each aircraft type three empty weights are quoted - the complete aircraft must not be heavier than the greatest weight, and the CG must lie within a range measured from the rear of the firewall datum; for example

For weight not exceeding	2300lb	2400lb	2500lb	
CG must lie not further forward than	30"	32"	34"	behind the datum
nor further back than	38"	36"	35"	behind the datum

Linear interpolation is used for intermediate weights.

There is also a note that the weight, if the aircraft is built only of PS components, is so much, for example, 2350lb. This makes it a little easier to work out all-up weight. The CG position for this weight is also shown. Hence, by substituting AS weights and moments as necessary for PS ones, we can calculate the effect on weight and CG.

Master Components File There is a master component file 'MCF', also made up of A4 sheets in binders, arranged in component number order, whether a Bluebird number (B ...) or an outside supplier's. Each sheet has the following details:

Component code

Short description and blueprint number, manufac. spec., technical data reference

Dimensions and weight (not distance from datum, since it may be fitted in several places)

Power requirements, environmental limits, etc.

Source (name and address of vendor)

Price last negotiated and delivery punctuality

Contract No. if any

Standard cost price

Quantity usually bought

Bin location in stores; target holding (can be 0)

Stock control information, viz

    BOH, expected receipts by date,
    expected usage, by date

Lead time for obtaining

Inspection procedure

Where used (cross-references to CS)

    e.g. CS100.2.3.
          CS100.8.9.
          CS160.7.7.

(Blueprints and manufacturing specs. for Bluebird parts are held separately.)

Method of Use

In Manufacture When an order for n Bxxxs is received by the factory from sales, for delivery by dates $d_1, d_2, \ldots, d_n$, the factory will accept it only if the earliest date allows for the longest delivery + manufacture lead time, since once or twice they have been criticised for late completion. This does mean that they lose orders: if they could quote and meet the nominal delivery time they could sell 30% more.

Preliminary On receipt of the orders they work down the appropriate S for each order, level by level, cross-referring to the MCF. Stocks of components are not kept, although there may be odd components remaining as a result of cancelled orders or over-ordering. Any components needed by the new order are earmarked if they are in stock or due in stock at the required time; normally, however, they are specially ordered. PS components are used for preference. The MCF is amended in pencil to show the planned allocation, and purchase orders and shop manufacture orders sent out.

Production Control expects confirmation of these orders in a few days, and may telephone if necessary to find out if all is well. If it seems the supplier will not be able to cope, it will be necessary to raise an AS order. If this happens PC has to run through the S and limitations to ensure that weights, CG and logical relationships are alright. It may even mean changing other orders to AS (this is one source of extra components being held in stock).

<u>Just Before Assembly Starts</u> About a week before the first set of components for the lowest level assemblies is due into the shop, PC raises assembly works orders by referring to the S again, the MCF, and taking dyeline copies or Bandas of the works instructions. They put a due date on these, and pass them to the shop.

<u>On the Due Day</u> (for starting) The shop foreman will check that all the components specified on the shop order are present; if not he consults PC, who will carry out a search for spare components, possibly AS, which can be used instead. If so, some rapid recalculations will be necessary: it may mean recalling many other shop orders and altering them.

On occasion, PC will 'rob' another set of components for another aircraft order, or borrow from a spares holding which properly speaking is reserved for urgent repairs to customers' aircraft. (Only a selection of such spares are held - mostly fibreglass fabrications and nuts and bolts. Other components would be reordered from suppliers.)

The same situation can arise when a component is ruined during work and there is no spare. The shop foreman has been known to take orders just arrived for a later shop order, and use these on the present one. Later on there will be a problem.

Two days before the order is due out of the shop, PC rings up to see if all is well (a copy is kept of the shop orders in date of completion order). If the order looks like being delayed, they will first try to speed up, and if not possible, will warn the shops dependent on the results of the earlier operations, so they can replan their work-load.

<u>Modifications</u>

As well as PS-AS substitutions, Bluebird deals with modifications of three kinds

(A) Effective immediate: all aircraft in manufacture and

already manufactured to be retro-fitted. This is a 'panic' situation caused usually be an incident involving safety. This plays havoc with normal production and the works concentrates on modifying returned aircraft or building mod kits for despatch to owners. This is fairly rare, fortunately.

(B) Effective ASAP: complete any assemblies already started to old specifications, but all new ones to have the modification. This will usually mean some old components will be left over. These can sometimes be disassembled, and common subcomponents re-used. Bs occur once a week on average.

(C) As and when: use up old stocks, but for new orders, use the modification. Cs also occur once a week.

Actions: for A, delete the old component from all specifications, and promote an AS to take over if the old one was a PS. (Or introduce a competely new PS.) B, C as A, but the modification is marked with a target date for changeover.

Statistics

Maximum number of levels 4

Components, average per level, 8

Commonality, none at level 1 (between aircraft types) but at levels 2 to 4

50% of components are used on	1 assembly only
25%	2 assemblies
15%	3
5%	4
2%	5
3%	6 or more (maximum 200)
100%	2 approximate weighted average

Hence total number of different components in use

3 x 8 at level 1	24
3 x ($8^2 + 8^3 + 8^4$) at other levels, divided by an average of 2	7008
Total:	7032

47

<u>AS Components</u> For each of the above PS components there is an average of 0.5 AS components (range 0 to 10).

<u>Illegalities</u> There is an average of three illegality rules per structure of average 8 components, each referring to a minimum of two and a maximum of 10 components, average 4.

<u>Lead Times</u>

      Assembly, test, inspection: average 4 weeks
      per structure, occasionally (20%)   6 weeks

Hence total assembly time, if not held up by shortages

    Average: $4 \times (0.8 \times 4 + 0.2 \times 6)$ = 17.6 weeks
         Minimum   16 weeks
         Maximum   24 weeks

    Manufacturing/Purchasing
    components   10 weeks

Hence the best performance is 16 weeks, or if assembly takes maximum time, and there is a shortage at each level, (not detected until assembly due to start) it could be as much as

    $4 \times 10 + 24$ = 64 weeks

Normal planning is based on 25 weeks to 34 weeks from receipt of firm order. In 70% of occasions this is met, but through planning errors in 30% of cases this is exceeded. Of these 30%, 20% are still finished, 10% are cancelled by the customer and the components stripped down back to stock.

<u>Sales Per Annum</u>   <u>Revenue</u>

    20 B100s x 12 000   240 000
    15 B160s x 16 000   240 000
    10 B200s x 18 000   180 000
                                 £660 000

It is estimated that sales could be very much improved if deliveries were more reliable and/or faster – up to 50%, possibly

<u>Costs</u>

    Materials cost per a/c, nominal   40%) of sales
    Manufacturing, assembly, etc.   15%) price.

Financial summary – at nominal costs

```
Total cost of sales  55% x £660 000 = £363 000 per annum
Administration and indirect labour       50 000
Other overheads (factory etc.)          100 000
Gross Profit                            147 000
                                        ───────
Revenue as shown above                 £660 000
```

However, the nominal materials cost is increased in practice, through the use of AS components which on average cost 30% more

    25% use of AS parts where allowable

    (50% of cases) x 30% x £363 000 = £13 612 extra cost p.a.

YOUR BRIEF

Write a report, which indicates the opportunities for improvement in financial terms, by amending present procedures.

Design a computer-based system to secure these improvements, with

> file structures and contents,
>
> program functions and structure,
>
> clerical procedures,
>
> documents.

Using the computer of your choice, e.g. ICL, IBM, CTL, determine the hardware and software required, and the approximate load. Draw appropriate charts, layouts, etc.

The report is addressed to

> Production Control Manager,
>
> DP Manager (who has a computer of the kind and configuration you need, and plently of spare time on it. Present applications are sales and purchase ledger, and payroll),
>
> Accountant.

7
The Garden Ornaments Management Information System Project

In October 1978 the Chairman of Garden Ornaments was quite pleased with the computer department. Invoicing, stock control and payroll had been on it for two years now, admittedly without showing all the cost savings that had been hoped for, but without any catastrophic muddles such as had almost put Concrete Mushrooms out of business. The chairman and the financial controller discussed the next step, which they felt should be a major one for the following reasons.

(1) Further improvements to existing jobs on the computer would be of only marginal benefit.

(2) The activity which had the biggest influence on GO's turnover and profitability was that of general management. If the computer could raise the quality of decisions, even if only a little, the benefits would be much greater than could ever be obtained from clerical savings.

(3) The chairman had attended seminars on real-time and management information systems and had been impressed.

(4) The computer department were getting restive because, as they said, they were on second generation equipment and jobs, instead of third or fourth. They were getting out of date.

At the same time the chairman and the financial controller were determined that the project should be planned on a professional basis, using such techniques as Discounted Cash Flow. The Data Processing Manager was called in and told about their ideas for GOMIS - Garden Ornaments Management Information System. It would provide real-time cathode ray terminals for 20 senior GO managers (five in Scotland) connected to a data-base at the centre. Any question could be keyed in, and detailed or summarised facts sent back by the computer. It would also 'learn' by experience and recommend courses of action. In Phase 1 these would be advisory only; in Phase 2 managers would be expected to give formal reasons if ever they overrode the computer proposals. The DP manager went away feeling strangely nervous but very excited by GOMIS, to do a feasibility study (although in effect the go ahead had already been given).

After all, GO's turn-over was

$$£48m \text{ per annum}$$

and profits were

$$£4.8m, \text{ i.e. 10 per cent on turn-over (='profitability')}$$

Both turn-over and profitability must be affected at least 10% by the quality of management. If this could be improved with computer assistance by a factor of 25% turn-over would rise by

$$£48m \times 10\% \times 25\% \text{ i.e. } £1.2m \text{ to } £49.2m$$

Profitability would rise from 10% by

$$10\% \times 10\% \times 15\% \text{ i.e. } 0.25\%$$

Profits would rise, therefore, to

$$£49.2m \times 10.25\% \text{ i.e. } £5\,043\,000$$

an increase of

$$£243\,000 \text{ per annum, or } £240\,000 \text{ to the nearest } £10\,000$$

The DP manager found that he certainly needed this revenue to pay for the costs of the project, even after charging all his present computer and staff costs to his existing applications (so that only extra equipment and development were charged to GOMIS). When he confided his worries to the financial controller, the latter said that the £240 000 per annum was a conservative figure: that there would be other intangible benefits, and that the old man had the bit between his teeth, so (to quote the chairman's favourite pun) it was "all systems GO".

The IBL salesman, the DP manager, and a very bright young software expert worked on the feasibility study. The hardware was fairly easy to specify. There would be two central processors: one would be capable of providing a 'degraded' service only, if the main processor was down, and would be used at other times for off-line work.

They were less sure about the file structure and retrieval methods, however. The IBL salesman proposed the use of three packages whose detailed specification would be available by January 1st, 1980.

PHILO: a generalised file management package,

PHORK: a generalised forecasting and decision-making package,

K415 version E: a real-time operating system executive.

They could go and see an early release of PHORK working in Finland. Some programming work would be needed to tailor these packages together, but clearly it would be preferable to use them rather than write their own. The actual working programmes would be supplied free of charge, nine months after the specifications, that is, by 1st October, 1980.

FEASIBILITY STUDY

At this stage it looked as if equipment delivery would be the critical factor, and an early letter of intent was vital. The DP manager thought the chairman would willingly sign without even seeing the feasibility study, but preferred to get it finished first for his own peace of mind anyway. It contained

A general description of GOMIS

A sample 'conversation' between a manager and the system

Some very approximate design calculations justifying the configuration, based on 20 managers making 10 accesses a day, plus 25 000 other accesses for updating the files, making 25 200 per day

A critical path analysis of project development

A Gantt chart of activities and staff requirements (see Table 2)

A DCF/payback analysis (Table 1)

A note on assumptions made, e.g. re benefits

(written by IBL) Proposed hardware and software: cost details.

Table 1 shows that the project gives a good DCF return on expenditure: 20% plus over eight years. The DP manager felt a good deal happier after this analysis was complete.

During this period some talks and seminars were held in the presence of the 20 senior managers, or, more often, their deputies. Rather to the DP manager's surprise, there was

Table 1 GOMIS economic analysis   DCF = 20%

Year	Qtr	Equipment	Staff	Training	Benefits	Net income	Present day value	Cumulative income
1	1		10					
1	2		20					
1	3		20					
1	4		20			-70	-70	-70
2	1		20	5				
2	2	30	20	5				
2	3	20	20	5				
2	4	20	20	5		-170	-142	-212
3	1	20	20					
3	2	20	10					
3	3	20	10		60			
3	4	20	10		60	-10	-7	-219
4		80	40		240	120	70	-149
5		80	40		240	120	58	-91
6		80	40		240	120	49	-42
7		80	40		240	120	40	-2
8		80	40		240	120	33	31
9		80	40		240	120	28	59
10		80	40		240	120	23	82

Notes: £10 000 installation costs are included in the first quarters' equipment costs.
Training costs are estimated overtime payments and other expenses.
System is operational in 2nd quarter of 3rd year.
Staff can then be partly redeployed. Benefits begin a quarter later.
The effect of inflation on both costs and benefits is ignored.

no opposition: the feeling was 'You cannot stop progress'. However, when he tried to find out just what questions would be asked of the computer, how often each day, and how the answers were related to the decisions taken, he found the replies unsatisfying. None of them denied that they had need of large amounts of relevant, timely information, nor that their present performance was limited by lack of it, but in general they refused to be specific about what they wanted, and felt that they might ask absolutely any question at any time of day or night. When the DP manager said that the cost of the system could be drastically reduced if the period of computer availability were cut down (for example) to two four-hour periods out of the 24, so that the databank could be updated (in batch mode) between these periods, he was met by blank stares and insistence that the data must be 100% up-to-date, and continuously accessible, as the chairman had promised in his introduction to the seminar.

The DP manager decided that instead of attempting to learn directly from the senior managers what their functions were and what information at what frequency they needed to perform it, he would himself define their responsibilities, and deduce what information service they ought to need to meet them.

A letter of intent was placed and recruitment of systems analysts and programmers with specialist real-time knowledge began. GOMIS was under way, in January 1979.

PROGNOSIS

Will GOMIS succeed? It has been started in accordance with many accepted rules

Top management involvement and enthusiasm

A quantitative assessment of the costs and benefits

A plan for implementation

Application of the computer to the central problem of the business

What can go wrong?

Table 2

GANTT Chart. Activity (if activity 'y' has No. (x) shown by it 'x' must end before 'y' starts)

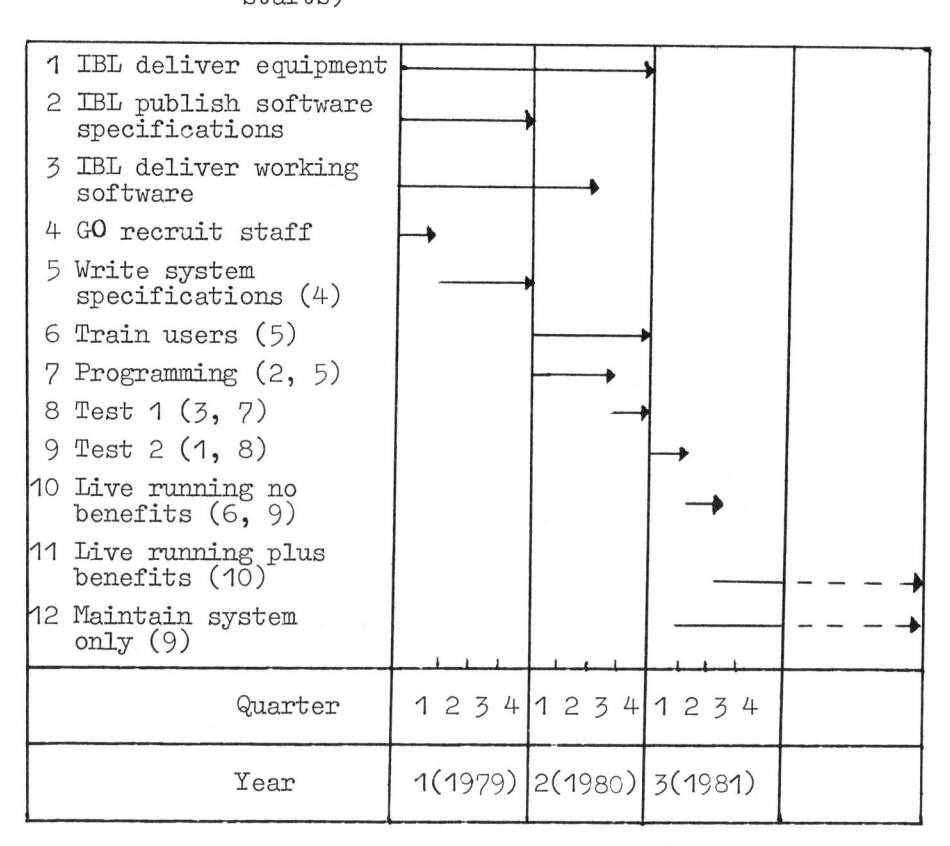

Activity	Timeline
1 IBL deliver equipment	
2 IBL publish software specifications	
3 IBL deliver working software	
4 GO recruit staff	
5 Write system specifications (4)	
6 Train users (5)	
7 Programming (2, 5)	
8 Test 1 (3, 7)	
9 Test 2 (1, 8)	
10 Live running no benefits (6, 9)	
11 Live running plus benefits (10)	
12 Maintain system only (9)	
Quarter	1 2 3 4 | 1 2 3 4 | 1 2 3 4
Year	1(1979) | 2(1980) | 3(1981)

EQUIPMENT

Delivery With IBL (or ICM for that matter), late delivery is not usually a serious problem. GOMIS requires equipment installation two years after date of order (see Table 2) which should be readily attainable with normal manufacturing cycle times of 12 to 18 months.

Cost A price increase will probably be matched by at least equal inflation in the benefits. However, even if it is not, a 10% increase, for example, means that the DCF rate of 20% plus has to apply for a period of nine years instead of eight: not a significant change.

Table 3

Revised GOMIS economic analysis - benefits delayed a year, DCF rate reduced to 10%

Year	Qtr	Costs Equipment	Staff	Training	Benefits	Net income	Present day value	Cumulative income
1	1		10					
1	2		20					
1	3		20					
1	4		20			-70	-70	-70
2	1		20					
2	2	30	20					
2	3	20	20				-137	-207
2	4	20	20			-150		
3	1	20	20	5				
3	2	20	20	5				
3	3	20	20	5			-148	-355
3	4	20	20	5		-180		
4	1	20	20					
4	2	20	10					
4	3	20	10		60		-8	-363
4	4	20	10		60	-10		
5		80	40		240	120	82	-281
6		80	40		240	120	74	-207
7		80	40		240	120	68	-139
8		80	40		240	120	61	-78
9		80	40		240	120	56	-22
10		80	40		240	120	51	29

Reliability It is not unusual for hardware to be significantly slower or less reliable than claimed. However, we can assume that GOMIS will have adequate safety margins built into the computer operating procedures, and the effect will be similar to that of a cost increase (although if the equipment is rented, part of the cost of unreliability can be recovered by reclaiming rental).

Software

This is quite a different kettle of fish. The delivery and performance of software are as critical today as those of hardware in the early days, and the decision to make use of the two packages PHILO and PHORK, and the real-time operating system executive K415 version E, is fraught with danger since none of these is operational.

The project timetable demands the prompt arrival of the software specifications on January 1st 1980, so that programming can start, and of the working software itself on October 1st, so that testing can start.

It is reasonable to suppose there is a significant chance that at least one of the three IBL projects may run into snags which will delay the issue of specifications or working programs and have an impact on GOMIS. (What would probably happen under these circumstances is the IBL would release partial versions on time, if possible, which would omit features that had been found tricky or extravagant to implement: or release a version that used a great deal of computer time and space, with more efficient ones later. This will mean that Garden Ornaments themselves will have to expend quite a lot of energy in just keeping up-to-date with the various unofficial and early versions of PHORK, etc. (If they don't their programme testing schedule will suffer.)

Just as a result of delays and inconveniences caused by late or incomplete issues of software, it is quite likely that the development of GOMIS could be delayed by a year. The effects of this on an economic analysis would be marked: the project would yield only 10% over 10 years instead of 20% plus over eight, on the assumption that user training costs could be postponed a year, but not those of the computer installation and the development staff. A revised version of the original economic analysis is shown in Table 3.

What precautions could be taken? The obvious advice would be: avoid having activities on or near the critical path of your project, if they are out of your control. This means either sticking to existing software (if necessary overcoming drawbacks by more or faster hardware) or by getting some degree of control over IBL, for example, by an agreement which gave the customer free use of the computer until the software passed some previously agreed acceptance test.

On a more cheerful note, the practice is on the increase of paying for software, from a programming house or from a manufacturer. This must be beneficial for the average user. He is motivated to use cheap, therefore simple, software, and not to hope that he can get highly sophisticated systems off the ground on a shoestring by using grandiose free software and minimal hardware. Also, since he pays the piper, he has far more control over the quality and promptness of tune he gets. Joint progress meetings and penalty clauses spring to mind.

Staff Recruitment

Three months were allowed for building up the system analysis team to a point where systems specifications could be started. This is certainly going to be a tight schedule, if one bears in mind the time required to advertise, interview, select, offer, get answer and wait for staff to work out notice with previous employers. But it would have been dangerous (even if it had been possible) to have built up the team earlier. If the project is still at - or has returned to! - the initial feasibility and approval stage, when only two or three people can usefully work on it, the new recruits will not be used when they arrive. The newcomers are well-qualified and mobile (facts just demonstrated by their recruitment), and the old staff are probably feeling unsettled by the newcomers. Unless they are all kept busy to a progressive, demanding schedule of work, both groups will leave.

When agonising re-appraisals occur, staff morale is a major problem. Is the DP manager to be quite frank and open about the difficulty? Or keep it from them and stick to the original plan, hoping that not too much of the detailed work being done will turn out to have been wasted?

At all events, this activity of staff recruitment is a critical one for the development of GOMIS, and any major hold-ups in development are likely to be compounded by the phenomenon described above: the impatience of computer staff.

System Design: Technical Aspects

Ideally the stages in a computer project are

    Feasibility study

    Economic evaluation

    Approval

    Systems design

    Implementation (programming, training, testing etc.)

But a much more common one is

    Outline idea

    Outline approval

    Systems design

    Major snag

    Belated feasibility study/economic evaluation

    Systems re-design

    Implementation (after possibly several more 'snakes and ladders' recursions from the discovery of snags back to the feasibility/economic evaluation stage).

Many DP managers go through life with an enormous guilt complex because their projects follow the latter course, when they had aimed for the former and believed the former to be desirable and practicable. This guilt is deserved if the project is a well-understood one whose costs, benefits, and techniques are well-known and safe. But a project like GOMIS is like snakes and ladders! GO deliberately chose a high-risk, high potential pay-off computer application. Admittedly, the DP manager and his staff can be criticised if the project keeps on returning to square one because of bad design calculations or over-optimism in the validity of untried techniques: but he and his bosses are much more to be blamed if they do not recognise the risks, and take into account in evaluating the project not just the most optimistic outcome (as was done in the original DCF) but also the others.

It is during the stage of systems design that most of the dangerous problems will crop up. (In an ambitious project, if problems do not crop up now, it is likely that this is because the work is not being done critically enough, and the undiscovered problems will arise after the equipment has arrived and while the programs are in a late stage of testing: a much more expensive business). The economic effect on GOMIS of delays (from whatever cause) in the development schedule has already been examined in the preceding paragraph 2 on software.

<u>Systems Design: User Aspects</u>

Part of the process of systems specification is the production of 'user manuals' which describe in layman's terms how GOMIS is to be operated, and the benefits that will be obtained from the computer's assistance. These manuals are to be agreed with the senior managers, and will then serve as the basis for the training programme due to start at the beginning of 1980.

It will be realised that GO's DP manager gave up trying to maintain communication with the senior managers after the first few unsatisfactory meetings, and worked out for himself what GOMIS should do for them. The submission of the 'user manuals' will be the first contact for a year. Now the managers will begin to realise that GOMIS is going to happen, and start to read the small print. They will not, of course, criticise the technical details of the system, but will very probably dispute

> the description of their own functions
>
> the benefits of computer assistance
>
> points of detail in the proposed new system.

They are very fearful of being put into a position where they are expected to operate a new system they do not understand, or to reduce their costs. They will therefore seize on any error of fact in order to prove that the new system is misconceived and they need no longer adopt it.

The DP manager will apologetically amend the manuals and very likely the system itself, hoping to meet the objections. However, he will not succeed, since at bottom the managers are rejecting not points of detail, but the whole idea of a change in their functions which, they feel, is being imposed on them. In the end, perhaps, it will be imposed on them by the chairman's authority. The managers will then operate the system, hoping to prove it is no good.

In any computer project the DP manager must keep up a dialogue with his 'customers' - the general managers. Where the project is a 'management information system' this is even more vital. If managers are given new information to a new time-cycle their jobs are inevitably changed. Computer people sometimes feel they can do nothing but good if they provide managers with better, faster information. In fact they are disturbing the ecology of the management structure much as a farmer disturbs the ecology of his fields with indiscriminate use of 'beneficial' chemicals.

WHAT IS TO BE DONE?

What advice could we offer the DP manager when he begins to

see the dangerous course on which he has embarked? Cynics may say - look for another job, even quoting the GOMIS project as a feather in his cap!

But he has a duty to GO, and indeed to his profession as a data processing man. He must

(1) Bring it home to his top management that GOMIS will probably be more important for good or ill than any other single venture yet undertaken by the company, and that they must devote time and effort to understand and direct the evaluation, planning, and implementation of GOMIS. After all, if it were a new factory in Northern Ireland they were considering, they would expect to get involved in questions of floor loading, tax implications, transport costs, etc.. With GOMIS they must get involved with 'response times', 'file security', etc.

(2) Even more important, there must be an agreed statement of GOMIS' objectives, and agreed criteria of success, preferably expressed in money or some other quantitative terms. Any arguments about the development of GOMIS can then be resolved by the touch-stone. What action increases the chances of achieving the agreed objectives? Of course these objectives will probably not remain static, and could well have been altered by the time GOMIS is running, because business or computer technology has changed. But such alterations will have to be deliberate and subscribed to by all concerned.

(3) Keep a cheerful face. Whatever his private views, a defeatist attitude will harm his own interests and GO's. Could it even be that the chairman's optimistic view of GOMIS, based on his intuition and courage to take risks, is right, and the DP manager is being narrow-minded?

(4) Offer an alternative. He should give his management the facts and persuade them to make an informed decision on GOMIS: but he must anticipate the question: What do we do if we reject GOMIS? 'Nothing' is not a good enough answer. It would be better to be in a position to suggest, for example, the introduction of a fairly orthodox budgetary control system coupled with management by objectives. This would require a modest increase in his staff and equipment, and would be a much 'safer' line of development.

(5) Keep his options open. While GOMIS is in the melting

pot he must postpone as many irrevocable decisions as possible, and so minimise the investment risk. Easier said than done! He will doubtless have some tricky interviews with manufacturers and staff.

After all the misgivings and reappraisals, perhaps GOMIS or a new version of it, will finally be approved again, but this time understood and supported by everyone affected. And it will succeed.

YOUR BRIEF

This case study is intended more as a discussion paper then a formal problem. You are invited to have a seminar/tutorial on GOMIS, or to draft a letter from the DP manager to the chairman, alerting him to the dangers, and proposing actions.

8
Brighton Rock

The old fashioned way of making seaside rock is to build a short fat cylinder of rock out of many thinner ones of different colours, and then to roll the big cylinder so that it becomes long, thin and compact: it is then chopped into appropriate lengths for sale.

Brighton Rock uses a patented machine which extrudes the final product in a single go. Hoppers of coloured sugar (kept molten by heaters) release their contents to pumps which force the sugar down small pipes, which are combined to form a 'dot matrix' to display the required patterns and/or words as shown.

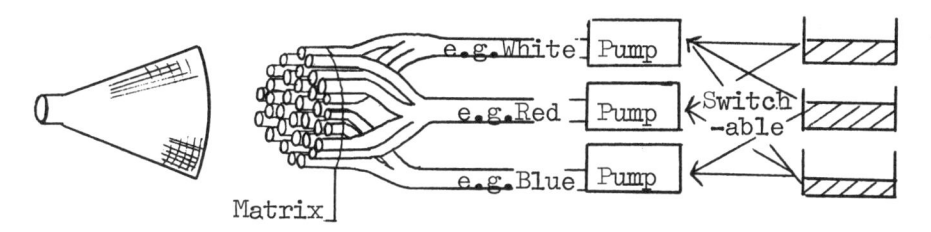

There is a different matrix for each different rock, for example

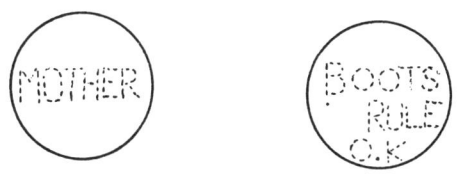

The extruded sugar is forced through a funnel to compact it.

The kinds of rock produced are

A_1 White, red, green A_2 White, red, blue

B_1 White, red, green B_2 White, red, blue

C_1	Red, green	C_2	Red, blue
D_1	White, blue, green	D_2	White, green, blue
E_1	White, red	E_2	Red, white
F_1	Blue, green	F_2	Green, blue

Hence to produce C_1 we need to fit the C matrix to the machine, and reload hopper 1 with red sugar, hopper 2 with green (if they are not already loaded from the previous job).

Work Content

The machine runs automatically when set up, but setting-up takes time

to clean a hopper of sugar and refill with another colour	30 min
to connect a hopper to another pump	5 min
to change a matrix	15 min
to start up the machine	5 min

(More sugar can be added without stopping the machine.)

Load

The machine runs for two shifts a day, five days a week. The procedure is for rock orders to be entered in a rack **in order of receipt** as follows (30 orders in total)

Order No.	Due	Quantity	Code
51	5 Nov 80	2000	F_1
52	3 Nov 80	3000	E_2
53	10 Nov 80	2500	A_2
⋮	⋮	⋮	⋮

The production strategy is simply to wait for a machine to become free, then set it up to do the lowest numbered order in the rack. In many instances this strategy results in a great deal of set up time, for example, if the machine was doing A_2 and had to take on order 51 next, the Blue sugar tank could be left (but changed to another pump), but another tank would have to be changed over to Green sugar: the total change-over time will be

$$5 \text{ min} + 30 \text{ min} + 15 \text{ min} + 5 \text{ min} = 55 \text{ min}$$

(The Red hopper would not be used.) If, however, the machine had been allowed to continue with A_2, it could complete order 53 without stopping - but this contravenes the existing First In First Out rule.

Actual running time is 1 min per ten units of quantity, thus order 51 will take 200 min, apart from the set up.

Productivity

As a rule of thumb, Brighton Rock allows 70 min set-up for each order, and the average duration is 250 min, hence the 'system' can handle roughly

$$\frac{60 \times 5 \times 2 \times 8}{250 + 70} = 15 \text{ orders a week}$$

Pumps, hoppers, and matrices break down from time to time, in which case some product types cannot be produced.

Many more orders are received than can be handled. Each day the rack is pruned by removing any orders which would now be delivered late, for which the customer will not accept a delay. (If he will, the order is left in the rack and given a new date.)

YOUR BRIEF

Propose short and long term improvements to the FIFO scheduling system to show what benefits in terms of increased productivity could be gained. Recommend appropriate O & M and information systems support, and show costs involved, bearing in mind that the added value of the average order is £200.

Consider particularly the incremental value of increasing sophistication and rapid response, and of more detailed investigations and simulations.

You may assume you are a member of a systems consultancy team sent down from the Head Office of the holding group (Universal Foods) at the request of Brighton Rock. An internal charge of £50 per man-day is made for the use of your services.

9
Camelot Ice Cream

Camelot makes high quality ice-cream packs selling retail at 75p, mostly in quality restaurants and hotels. It keeps for 60 days in deep-freeze depots, lorries, and vans but for only 10 days in hotels and restaurant refrigerators, where it is kept at a higher temperature ready for consumption. It contains real cream and fruit. Packs which have been in stock 59 days (at depots or in central warehouse or in vans) are allowed 10 days in customer refrigerators even if sold on the 59th day. None should be sold on the 60th or later days.

To protect its high quality image, Camelot requires customers to return packs if unsold after 10 days and refunds the wholesale price, 40p, by cheque from head office. Such returns, and any packs in deep freeze longer than 60 days, are sold for animal feed at 1p each.

Camelot has a factory in Birmingham and a head office. Two large refrigerated lorries A and B deliver to depots as follows

Depot	Cost of depot space £ p.a.	Delivery day(s)	Lorry
London	20 000	Mon, Fri	A, A
Manchester	15 000	Tues	B
Birmingham	15 000	Wed	A
Southampton	10 000	Thurs	B
Exeter	5 000	Fri	B
Llandudno	5 000	Mon	B
Brighton	10 000	Thurs	A
	80 000		

There is a clerk and storeman at each depot (two storemen in London). The clerk's duties are as follows.

Replenishment Each depot is replenished to the maximum space available. The clerk sends a telex to Birmingham head office by 5.30 p.m. on the day before the lorry comes. This telex shows

Balance brought forward, i.e. as last telex	a
Received by lorry since then	b
Issued to salesmen	c
Balance carried forward: a + b - c	d
Physical stock check: time-expired packs	e
Packs 40-59 days old	f
20-39	g
under 20	h
Total physically stock-checked today: e+...+h	i
Discrepancy if any between d and i	j
Depot capacity	k
Replenishment quantity requested: k - i + e	l

(Lorry is to return e to Birmingham for disposal)

Packs have date of manufacture clearly stamped on each and depot bins are colour coded to assist stock-checks.

Salesmen Sales are made by van salesmen on commission. They have distinctive deep-freeze vans with no capacity problems, and have a great deal of freedom in their handling of their territory. They come into the depot each morning at 8 a.m. and the depot clerk carries out the following procedure.

Cash and cheques: receive these from the salesman. He will have collected them from medium/small customers (M/S). Book into cash book and credit salesman's 'pack book' (PB) with the equivalent number of packs at 40p.

Time-expired packs: exchange any returned by customers or the salesman from his own van stock, for fresher ones, and put old ones in time-expired stock for collection. No paperwork except that the salesman should give the clerk forms for refund (RF) signed by him and the customer, to be sent to Birmingham where cheques will be sent to the customer.

Get Delivery Acknowledgement Forms (DA), signed by large customers (L). Credit salesman's PB with the equivalent number of packs and send DAs to head office for billing L customers.

Issue packs required by salesman and debit his PB: record issue in depot stock book. The oldest packs are issued first.

If the total number of packs shown in the balance on the PB is greater than four times his van's capacity, report this to head office.

Salesman:.....		Van:.....	PB Limit:	8000 packs	
Date	Packs b/f	Less DAs	Less cash, etc. divided by 40p	Plus issued	c/f
April 2	5000	1000	2500	3000	4500
April 3	4500				
etc.					

Example of PB

The packs b/f are the salesman's responsibility. For example, if he leaves the company, his terminal payment is reduced by the value of the packs on his PB, if not returned in full. If his PB shows more than four times his van capacity, it is probable that he has not collected DAs or payment from customers promptly.

Work-loads in Depots

From 8 a.m. until 9 a.m. the clerk and storeman (storemen in London) are busy dealing with salesmen, and loading vans.

The clerk spends another hour a day on paper-work. This can be scheduled at his discretion. The storeman (men) spends two hours a day on checking packs, shifting them into time-coded bins, etc.

The evening(s) before the replenishment lorry comes, the clerk spends from 2 p.m. until 5 p.m. doing a physical stock-check with the storeman(men) and from 5 to 5.30 sending off the telex.

On the replenishment day, the clerk and storeman(men) spend the afternoon (2 p.m. to 5.30) on moving packs from the lorry into store, and returning time-expired packs.

The rest of the time they have nothing to do.

Informally the salesmen arrange to stagger their time to arrive at the depot between 8 and 9 a.m. to avoid some wasted time waiting for the clerk.

The Salesman's Day

He has a defined territory which may include an L customer, although many salesmen do not have an L customer.

He visits the depot, disposes of cash, cheques, DAs and returns, picks up additional packs, and plans his itinerary. This will certainly include the large customer if any, and will also include those M customers not recently visited, and a selection of S customers and new prospects, typically

        1 L
        5 M (out of, say, 15 in his territory)
        2 S

In an eight-hour working day this means driving and visiting time per customer is about 1 hour. If he spends too long on planning his itinerary he loses time that could be spent on a call. Sometimes he calls at the customer's phoned-in request. At the L customer he delivers new packs and gets a signed DA. Rarely, he has to pick up time-expired packs. Since he calls daily it is unlikely this will happen often because he will reduce deliveries if the customer's stock has not gone down much. Returns are covered with an RF.

At the M customer he collects cash or cheque for the last delivery, gives a receipt, delivers further supplies, and picks up returns, which are covered by an RF. If no payment is forthcoming it is up to him to press for payment and suspend further supply.

At the S customer (small or prospective) he will urge the buyer to take some Camelot packs and at his own discretion could ask for payment in advance, pointing out the refund system.

After the end of the day he will return home. Camelot supplies an adapter to let him keep his deep-freezer running off his house mains while the van is garaged.

Salesmen do not work week-ends, and in fact nobody in Camelot does. This means that Monday is a heavy day for both sales and returns, since restaurants and hotels are particularly busy at week-ends.

Head Office

The factory is here, and so is the administration, marketing, and distribution headquarters.

There is a factory stock deep-freeze which buffers production and the lorry distribution to depots. It is aimed to keep this stock-room three-quarters full; if it drops to half-full, overtime is worked. If it reaches capacity, the factory is put on short-time.

The sales management is centralised at Birmingham, and the sales manager visits depots to sort out any personnel problems with salesmen as necessary, including local hiring and firing.

The company philosophy has been to simplify procedures and leave responsibility to the salesman. Stock control and production are run on very basic 'max/min' lines. No market research is done because the information is not available. However, some very expensive prestige advertising is done through TV and the Sunday papers to establish Camelot as 'up-market', although little is known about the actual purchasers.

The Accounts department send out bills to the L customers monthly, using an orthodox key-board accounting machine and one clerk. Refunds for returns are much more numerous (3300 packs returned per working day from 330 customers) and require three clerks to organise. The chief accountant also has a control section which validates the cash and cheque flows from depots. A distribution department of two clerks actions the depot telexes and organises lorry movements.

Statistics

Depot	No of Salesmen	No. of Customers L	M	S
London	15	12	200	100
Manchester	10	6	100	25
Birmingham	10	6	100	30
Southampton	10	5	50	25
Exeter	5	3	15	5
Llandudno	5	3	15	5
Brighton	10	5	20	10
	65	40	500	200

Days in working year: 52 x 5 = 260 .(relief salesmen and depot staff taken on at holiday time, but week-ends not worked).

Per working day average packs delivered by customer size

	L	M	S	Total
Customer visits	200	50	10	260
Average returns	10	5	2	17
Total delivered	8,000	25,000	2,000	35,000
Total returned	400	2,500	400	3,300

Revenue per day at 40p (wholesale price) per pack delivered not returned, and 1p per pack returned

£12,680 + £33 = £12,713

Revenue per annum: 260 x £12,713 =	£3 305 380
Less marginal cost per pack 10p	910 000−
Less commission at 1p on net sales	82 420−
Contribution to profit and fixed costs	2 312 960
Fixed costs: 65 salesmen basic salary	260 000
Fixed factory costs and wages	305 000
Central accounting and admin.	50 000
Lorries and drivers (2)	20 000
Advertising	187 000
Depot costs: space and electricity etc.	80 000
storemen and clerks	30 000
Management	50 000
Salesmens' special vans and expenses	65 000
Profit	1,265,960
Contribution to profit and fixed costs:	2,312,960

Opportunities and Problems

Fraud It is suspected that some packs which are returned for despatch to animal feed suppliers, somehow get into the hands of salesmen, who then return them on behalf of 'customers' and share out the refund cheque with the bogus customer. This could be reduced if customers or salesmen with an abnormally high return to delivery ratio were reported: potential reduction in refunds of £15,000 per annum. The practice could be completely eliminated by checking for duplicate pack numbers occurring on refund requests within a period of six months, and save a further £15,000, total £30,000 p.a.

Over-selling Some salesmen may have a habit of delivering excessive quantities. This causes losses to Camelot and reduces the salesman's own efficiency since he has to collect the returns. If salesmen could be warned of customers with whom they do business who return more than the average number of packs, the daily average packs returned could be reduced to 2,500 and deliveries increased to 37,500.

Marketing Information The £187,000 annual bill for advertising could be reduced (or an equivalent amount of extra contribution brought in) by £50.000 p.a. if the market could be better defined, by a report not necessarily more frequent than every three months, showing all customers' names, type (hotel, restaurant, other), and deliveries and returns in that period.

Salesman Efficiency If salesmen's itineraries could be worked out for them, by consideration of the customers on whom they should call that day, and the best route, they could fit an additional call in per day so that the total number of customers in each category could be increased by 15% without extra sales staff.

Depot Efficiency A study has shown that the space and electricity requirements of depots could be reduced by 20% if replenishment were worked out on a simple exponential smoothing system instead of max/min. This could be done with the help of a simple form and calculation sheet by the clerks, using the existing stock records.

Customer Satisfaction Sales are lost because customers who do not over-order sometimes run out before the salesman calls again. This happens mostly with M customers. It is possible to place a large red tag on the pack at the back of the customer's refrigerator with a phone number for him to ring to ask for a special visit the next day to replenish supplies. However, the present system does not lend itself to accepting such phone messages since there is no one in after 5.30 p.m. and the following morning the clerk is busy sorting out the salesmen. If such phone messages could be handled, M customers sales would increase by 15%.

Alternatively, if an information system had the complete ordering pattern of customers day by day, it could predict by trend analysis who may be worth calling on, automatically. This could increase M sales by 10% if no phone-in system were operating, and 5% if it were.

Bad Debts The profit figure of £1,265,960 is reduced in practice by about £100,000 which represents unpaid debts mostly from M and S customers. If the indebtedness of all customers were known within a month of the last transaction, the expenses of bad debts could be cut by 75%.

YOUR BRIEF

You are members of a team of consultants specialising in the design of computer-based systems to support sales order processing, distribution, and accounting. Camelot has asked you to suggest ways of tackling the above problems and taking the opportunities for improving Camelot's business, supported by technical and financial analysis, in the form of a Survey Report.

10
Vagabond Car Hire

VCH hires cars out from its eight offices in

>   Central London (Victoria)
>   North London (Watford)
>   South London (Croydon)
>   Heathrow
>   Birmingham
>   Manchester
>   Bristol
>   Glasgow

They use two types of car and charge rates

	Escort	Cortina
Per mile	4p	6p
Per day	£10	£14

(Petrol is also charged for.) Service is 7 days a week. Hire periods (not hirings) average out at: 50% one day, 25% two days, 25% over two days (average four), overall average two days.

10 cars of each type operate from each office (total 160) and there is a utilisation on average of 70% hire on each type, and when hired, average miles are 50. Hence revenue per day is

>   70% x 160 x (£12 + (5p x 50)) = £1624

This gives a total revenue p.a. of £592 760 against costs as follows

8 × (£6000 rents + 3 × £6000 staff)	£192 000
Depreciation and maintenance on cars 160 × £1000 p.a.	160 000
Head Office (at Victoria)	100 000
Total Cost	£452 000
Operating Surplus	£140 760 p.a.

The Wanderlust Project

The big rental companies allow hirers to return their cars to any office. VCH currently insists they bring them back to the office where they were hired, but is thinking about changing this, and a scheme called Wanderlust has been mooted. If hirers could return cars at any office, demand would increase by 25%, and average miles by 10%. However, it could happen that an office could become short of cars and so have to turn away hirers. At present each office expects to have six cars out of 20 still not hired by the afternoon. By midnight all the one-day cars (seven), half the two-day cars ($\frac{1}{2} \times 3\frac{1}{2} = 1\frac{3}{4}$) and a quarter of the longer period cars ($\frac{1}{4} \times 3\frac{1}{2} = \frac{7}{8}$) will have been returned = 9.625 cars. So next morning they will have 15.625 cars (including the six unhired) available, of which seven are hired for one day, $1\frac{3}{4}$ for two, and $\frac{7}{8}$ for more than two days, leaving six unhired, and so on. (In practice the demand will follow a Poisson pattern.) But under the Wanderlust scheme 25% more hirings are expected, providing more revenue per day

160 × 70% × 125% × (£12 + 5p × 110% × 50) = £2065

This means an annual revenue of £753 725 and although maintenance and depreciation have to be increased by £20 000 p.a. it increases the operating surplus to £281 725, that is, about doubled.

However, there are three snags

(1) Will the cars get out of position, so that some offices will have too many and some too few? This could mean a loss of business (or a need to have cars driven or transported back to their base, at extra cost) since the average level of six free cars is sometimes run down to zero even now.

(2) A new paperwork system will be needed so that the office where the car is returned can bill the customer correctly. At present the office staff

fill in a duplicate sheet at time of hiring and on the car's return, enter the miles, and calculate the total due, and give the hirer the balance of his deposit. Under Wanderlust, the hirer cannot be trusted to return his copy of the hire agreement, so the office of return must be able to have access to the information on the sheet made out at the office of departure.

(3) To make best use of the fleet of cars, it is proposed to let hirers have Cortinas at Escort prices if all the Escorts have gone. However, this arrangement could lead to fraud, in that office staff could charge hirers of Cortinas the Cortina price, then replace the sheets with sheets showing 'Cortina but Escort price' and pocket the difference. (This would obviously be easier for non-Wanderlust hirings.)

One possible approach to introduce Wanderlust safely would be to charge a premium on cars not returned to the office of hire, and progressively reduce the premium if the increased volume of hirings warranted it, and the snags were not too great. However, even this 'sticking a toe in' approach would demand the appropriate systems to be ready in advance.

YOUR BRIEF

You are asked to prepare a report for the Board of Vagabond Cars advising them of what computer systems could do

(1) to evaluate the economics of various possible strategies on the Wanderlust theme,

(2) to operate Wanderlust with special attention to

    (a) smooth, efficient paperwork

    (b) security from fraud

    (c) resistance to equipment/hire failure

    (d) management control information.

11
Canine Eugenics

1 <u>Your brief</u>: You are members of a software house, and a salesman from your office has just come back with an outline project following a visit to a company called Canine Eugenics. You are asked to turn his notes - which follow - into a proposal for the installation of a computer-based system.

This proposal should contain:

    Overall description of proposed new system

    Flow charts of information handling

    File and record descriptions

    Access methods

    Processing modules (program runs or subprograms with logic block diagrams or decision tables etc.)

    Form designs

    Notes on clerical and computer operating procedures

You may find it best to postpone making final decisions on many of these until you have had the appropriate lectures or discussions in tutorial.

2 <u>The problem</u>. Canine Eugenics Inc. (CEI) runs a "dogs' dating service". Applicants (people, not dogs) send in forms cut out of newspapers like the All American Hound Journal, giving the following information:

    Owner's name and reference code if a previous CEI customer (CEIC);

    Address if not CEIC.

    Dog reference code if CEIC and this dog previously advised, otherwise dog breed, name, date of birth, sex;

"Vital statistics" -

    up to 10 measurements in inches

        3 colour codes

        3 coat description codes   } (one digit)

        5 other miscellaneous codes

"Show successes" - show code (two digits and date)

                placement (one digit) for up to 5 shows.

CEIc customers need not fill in anything except their owner's reference and dog reference as a minimum, but may add new dogs or information about existing dogs.

Old customers need not send money but new ones should: $10 initial registration and one free "date": $5 per date thereafter.

At present, when the applications are received, file cards are made out for new customers and dogs, arranged by breed:

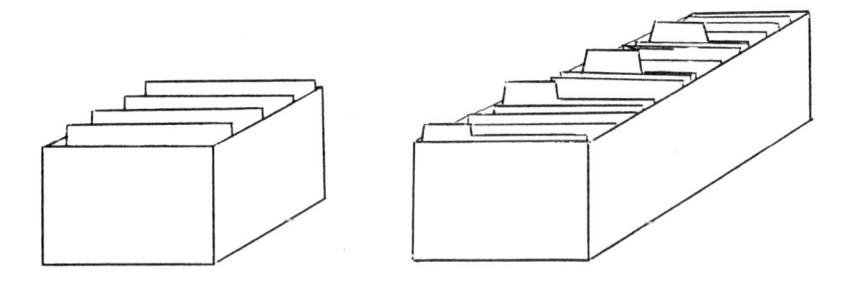

There is an index of customers in surname order giving registration numbers so that a file card can be pulled if the name is known:

```
ADAMS 7894
ADAMS 2467
BATES 9955
```

and a breed master sheet, showing:

> Breed name
>
> Statistics on show winners and their "Vital Statistics"
>
> Formula for calculating likely success of a puppy whose parents' "Vital Statistics" have given values.

Dog cards and customer cards are cross referenced to one another, by registration number.

3   Current procedure.

Daily: Open applications, tear up applications without money or registration number. Make out new cards and enter register (index) for new customers with money. For old customers, if no money recorded as being outstanding on their card, write down "$5" on the card, record any changes to owner or dog cards: create new dog card if necessary.

Whenever a card is created or changed, date stamp it. Old customers with money unpaid: send a standard letter asking for $5 for last time, and asking for a resubmission of application: tear up old application. If letters include money for previous occasions: find customer card and strike out "$5". Put coloured marker on all dog cards to be dealt with this afternoon.

In the afternoon: go through breed by breed taking out dog cards with coloured markers: consult breed master sheet and dog "Vital Statistics" to obtain best match out of other dogs in same breed: must be opposite sex (blue cards for dogs, pink for bitches), and try to match with a dog also flagged for attention this afternoon: temporarily remove pairs of dog cards as they are matched. At the end of the matching process, fill in a duplicate "Match Advice" to be sent to the owners of the two dogs, giving name and address of the other owner. If $5 had been written on a card, stamp "Payment $5 due" on the corresponding Match Advice.

Unmatched dog cards are put back in the file along with the matched ones, but still having their coloured markers. A sequence of colours is used each day, so that it can easily be seen if there are any carried forward dog cards, which must be given priority.

Note that a dog on file may get an unsolicited Match Advice and his owner will not be charged for it.

Monthly – Go through Customer cards and remove any with no activity for six months (by date stamp). Remove corresponding dog cards. Remove index entry. (One day we will need to re-use registration numbers, but we allow six digits so it won't be for a long time).

Go through breed by breed and update the Breed master sheet with the results of the latest shows,

and recalculate the Magic Success Formula Unique
to Canine Eugenics.

4   Statistics:

   Number of customers on file: 100 000

   Dogs per customer 1 - 10, average 2.5

   Applications per day   1 000

The computer you are to use has a card reader, Line printer,
tapes, discs, and can have 'mark sense' readers attached.
DO NOT WRITE ANY PROGRAMS. But your work should be sufficient for a programmer to use to write them and get the
system going. Use tape or disc or other methods as you wish.
Give reasons for your choice.

12
Banzai Pumps

1. __Introduction__. You are a systems analyst with a consultancy. A senior consultant has had an interview with the Managing Director of Banzai (UK) Ltd. and obtained information about this company's operations. You have been asked to study his notes and design a computer-based system which he can show to Banzai as evidence that it will be worth their commissioning a detailed study. Your task is thus to prepare an internal report.

It need not therefore include detailed costing (this will be done in the study) but should show

    Probable type and location of computer/bureau

    Types of peripherals needed including terminals, microfilm, etc.

    Allocation of main files to tape and/or disc: file structures and access methods

    Main applications planned

    Mode of operation in various parts of Banzai

You may also include any ideas you may have on what should be included in the study to be negotiated with Banzai, but try to produce the whole report inside two working days.

Banzai (UK) Ltd. markets 15 models of industrial pumps, ranging from £1 to £500, undiscounted price, average £10. Customers usually buy the pumps for inclusion in their own products. Banzai has a national headquarters and warehouse (NHW) at Watford, and distribution centres (DCs) there and at Bristol, Halesowen, Manchester and Glasgow. Orders are dealt with by the DC in whose territory the delivery is to be made. If a customer contacts NHW or the wrong DC, the order is passed to the right DC. Stock replenishment and distribution systems should be ignored.

The DCs use automatic typewriters with edge-punched cards to prepare despatch notes. The cards (giving the delivery address, customer, and invoice address) are prepared by NHW only, at the time the customer is first accepted.

Copies of the despatch notes are sent to NHW where they trigger off the sales accounting process, which is manual, and rather complicated.

Banzai deals with about 200 'special' customers which have the following typical structure

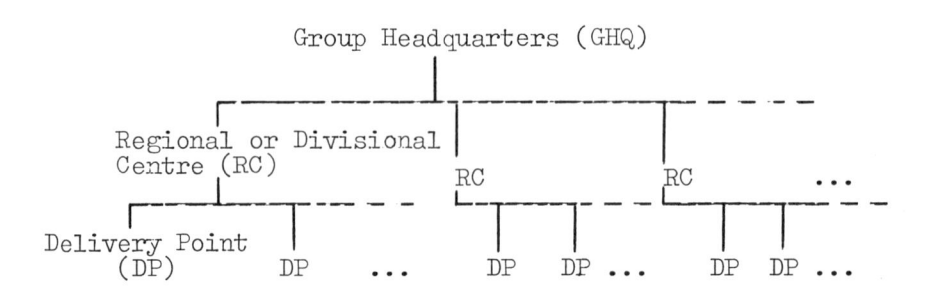

```
                    Group Headquarters (GHQ)
                              |
          ┌───────────────────┼───────────────┐ ─ ─ ─ ─
   Regional or Divisional     |               |
      Centre (RC)             RC              RC        ...
   ┌──────┴──────┐ ─ ─ ─   ┌──┴──┐         ┌──┴──┐
Delivery Point   |         |     |         |     |
      (DP)      DP   ...   DP    DP  ...   DP    DP   ...
```

A contract is worked out with GHQ every year specifying the likely quantity of each type of pump to be delivered per quarter, the price if actual deliveries match this estimate, and the price applicable for deliveries exceeding or falling short of the figure. Similarly a standard delivery charge is negotiated to the DPs. During the life of the contract, Banzai delivers pumps as ordered by the DPs and invoices the RCs at the contract price monthly, showing the DPs and individual deliveries. At the end of each quarter Banzai totals up all the deliveries and sends GHQ an invoice if deliveries were below contract level, or a cheque if above. (It will be up to the customer GHQ to apportion this adjustment among its RCs if it so wishes.)

Each GHQ has 2 to 20 RCs, average 5. Each RC has 1 to 10 DPs, average 2. The customer account is dealt with by NHW, often requiring telephone consultations with the DPs involved in delivering to the customer DPs, for example, re customer queries.

A problem arises when a customer queries an invoice item, that is, a particular delivery is alleged to be defective or short. The RC may then postpone payment on all deliveries in its area until the matter is cleared, or make a part-payment. Banzai has persuaded some special customers to make out a standing order to pay the figure based on the contract rate of deliveries, in order to prevent this. Any remaining adjustments are then done with the quarterly settlement.

There are also 1,500 normal customers, each with 1 to 10 DPs, average 2 (thus they are like RCs). No contracts are placed, but orders are billed individually according to the standard price-list with quantity and total-order-value discounts. In addition a customer status discount is allowed if the volume of business warrants it.

For both normal and special customers, each DP orders about £50 on average from Banzai each month, that is, an average of 5 pumps of various kinds, with an order frequency of 0.5 average, min 0, max 8 per month. Based on the undiscounted price list, the monthly revenue is thus

£50 x $[$(200 specials x 5 RCs) + 1,500 normals$]$ x 2 DPs = £¼m

A year's warranty is given on Banzai pumps which are claimed to be the very best quality. (To quote from the Banzai Nippon company song, freely translated from the Japanese, "Other people's pumps make sound like two hands clapping. Banzai make sound like one hand clapping.") Accordingly a record card is held for every customer showing each pump supplied, with date, pump serial number, original/replaced-under-guarantee, end of warranty date, type, end-use specified in order (if known), fault if any, and action if any (for example, replaced under warranty). Banzai keeps these cards going indefinitely so they can trace back histories of customer complaints. They do find it hard, however, to obtain statistics for faults on a particular type, over particular periods or serial numbers, irrespective of customer.

It may readily be seen that a year's business involves 300,000 pumps being shipped and about 10,000 new record cards being added to the records.

Each customer record also contains copies of live and dead invoices, copy despatch notes, etc. Clerks allocate incoming payments to invoices and chase outstanding invoices by sending reminders to customers. The complicated discount and contract schemes cause many difficulties but Banzai owes its success in the United Kingdom to flexibility in commercial negotiations. In fact management are considering introducing contract schemes for 'normal', that is, two-level customers, and prompt payment discounts over and above the existing arrangements.

Further Objectives of Banzai

Sometimes customers go bankrupt leaving Banzai with bad debts. A sign of impending bankruptcy is an increase in number of invoice queries and slowness in bill paying. Banzai would like to have these cases detected so it can consider stopping supply; conversely, if a customer 'on stop' for this reason pays up to date, they want to know quickly and resume supplies. This means giving DCs up-to-date stop-lists.

Sales Analysis and Marketing

Banzai codes at the RC level for 'end-use', and tries to correlate customers. Thus if a customer makes engines, and uses 25 type-2 and 50 type-3 pumps per annum, while another customer also making engines uses 25 type-2 but only 25 type-3, is it worth investigating to see if more type-3s can be sold to him? Similarly price structures and pump faults are correlated by end-use, etc., so that Banzai can see where customers of similar businesses and purchase volumes nevertheless have very different price structures or fault histories.

In particular, when a contract is being negotiated, it is essential for Banzai to be able to see whether, if this price structure under consideration were also applied to existing customers of similar volume and end-use, the revenue would be adversely affected. Such a 'simulation' facility would need to be made available over a period of a day or two.

It would therefore appear that the customer file at Banzai has many kinds of information on it, and is used for (or should be used for) many different purposes each with different response time requirements. It also looks as if the rules for invoicing and sales accounting may keep changing. What possibilities are there in a computer approach, and what sort of investigation is needed to explore them?

13
The PNP Electronics Company

1. PNP is a high technology research and manufacturing company. Its operations are apparent from its operating statement for 1980:

Component purchases	4,487,168	Sales	18,558,800
Assembly, test etc.	5,622,500	Contract Research	1,000,000
Research	2,000,000		
Overheads	2,000,000		
Obsolescence	393,750		
Surplus	5,055,382		
	19,558,800		19,558,800

2. <u>Components</u>. PNP purchases 8,000 different low value (LV) items, average 1p
and 2,000 different high value (HV) ones, average £1

Adequate component stocks are held and problems of stockouts and obsolescence may be ignored for this case-study except in so far as Substitutions are concerned. (10.)

3. <u>Assembly process</u>. Assemblies are made up of components or components and assemblies. To assemble and test and package and assembly, is costed at £25, using PNP or subcontractor labour, and takes five days.

4. Assemblies are produced at three levels:

Level 1: they consist of an average of 100 low-level components, of which 80% are low value resistors etc. costed at 1p, and 20% are high value, valued at £1. The cost of assembly is £25, so the average cost per L1 assembly is:

$$(80\% \times 100 \times 1p) + (20\% \times 100 \times £1) + £25 = £45.80$$

Most are produced as components for Level 2 (L2) assemblies but about 12 per working day (260 per annum) are sold directly at £90. There are 250 different types of L2 assembly.

Level 2: consist of an average of 50 low-level components and five L1s, so they cost: $5 \times £45.80 +$

$$(80\% \times 50 \times 1p) + (20\% \times 50 \times £1) + £25 = £264.40$$

Most are produced as components for Level 3 assemblies but 13 are sold each WD for £600. There are 250 L2 types.

Level 3: consist of an average of 50 low-level components and five L2s, so they cost: $5 \times £264.40 +$

$$(80\% \times 50 \times 1p) + (20\% \times 50 \times £1) + £25 = £1357.40$$

25 are sold per WD at £2500. There are 500 types.

Summary

	£Sales (net of tax)	£Costs
Level 1	280 800	142 896
Level 2	2 028 000	893 672
Level 3	16 250 000	8 823 100
Total	18 558 800	9 859 668

Analysis of costs:

Low level components per WD (average cost 20.8p each)

$$25 \times (50 + 5(50 + 500)) = 70\ 000$$
$$13 \times (50 + 500) \qquad = 7\ 150$$
$$12 \times 100 \qquad\qquad\qquad = \underline{1\ 200}$$
$$\qquad\qquad\qquad\qquad\qquad 78\ 350$$

Total component cost = 78 350 × 20.8p × 260 = 4 237 168
Total labour cost = 865 × £25 × 260 =   5 622 500
                                                £9 859 668

Actual component purchases and assembly costs are higher than as shown, through scrap and obsolescence (see below). The figures shown in 1. above, include these extra costs: £250,00 per annum component scrap, £393,750 obsolescence (made up of £135,000 stock losses and the remainder is write-off of fixtures etc.).

8. Production Planning. A 4G computer is in use to implement the following system, intended to meet all customer orders in 13 days.

   L1 assemblies are built on receipt of either a Customer Order or a Works Order. Customer Orders are thus met in five days. Works Orders are generated by the 4G by exploring L2 Works and Customer Orders, using a Product Structure File.

   L2 assemblies are built on receipt of either a Customer Order or a Works Order, after it has been exploded as above to provide the necessary L1 assemblies to enable work to be started. Customer Orders are thus met in 10 days. Works Orders are generated by the 4G exploding L3 Works Orders using the Product Structure File.

L3 assemblies are ordered singly by customers and are delivered within 13 days if at all, by holding a stock of one each of the 500 types. As soon as one is sold, the 4G replaces it within 10 days by issuing a Works Order for one, and L2 and L1 Works Orders to supply the assemblies for it. Since 25 L3 items are sold per day, the average per type is .05. If an order is received on a particular day, it can only be met if there has been no issue in the last 10 days. The probability of this may be found as $.95^{10} = .6$, hence to average .05 $\underline{issues}$ a day,

$$\frac{.05}{.6} = .08 \underline{orders} \text{ are needed.}$$

This represents .08 x 500 = 40 orders a day over the range, giving 40 - 25 = 15 refusals. It is not quite as bad as this because in practice customers can sometimes be persuaded to accept a longer delivery time. However another 20% of L3s could be sold, and probably more (in that customers would be more certain of being supplied) if this problem was solved.

The reason why a simple increase in the stocks of L3 is an inadequate answer, is that this would make even worse an existing problem due to obsolescence.

9. <u>Obsolescence</u>. L3 items particularly are liable to sudden obsolescence as a result of lower-priced competitive products coming onto the market without warning. They have a "half-life" of a year: i.e. half the range of 500 are affected. At present the average stock is .6 for each type. The cost per write-off is £900, which represents the cost of cannibalising usable assemblies and components, and scrapping the rest. Increasing average stocks by one per type will thus incur £225,000 per annum, which has to be set against the additional 20% of contribution (20% x 6500 x (£2,500 - £1,357.40)) = £1,485,380. And this additional contribution requires rather more than an increase of one only in the average stock levels. Obsolescence of L1 and L2 assemblies is not a problem because no stocks are held. If they were, obsolescence-risk costs are £15 and £100 per unit stocked p.a..

10. <u>Substitutions</u>. On a given assembly, it may be possible to use a different component from the one specified if the one specified is unobtainable. However, a component (taken in this section to mean component or L1 or L2 assembly) may be a permissible substitute for another on one type of assembly but not on another, hence substitutes are defined in assembly specifications. One quarter of all components have one or more substitutes in the case of any assembly. (Maximum 10, average 2). This possibility is of value when a component is out of stock.

Currently a whole department costing £15,000 p.a. (included in overheads) is employed to check stock records of components when the 4G produces the explosions of Customer and Works Orders, and to alter the Works Orders so produced, by hand, where the computer has specified components not in stock. Currently no <u>assembly</u> components are stocked since all are made specifically to order, but if they were, it would be useful to take advantage of commonalities between them in the same way. Only about 500 out of 10,000 parts are out of stock at a given time. A list of such parts is available daily from the stores.

Substitution rules are shown on the assembly specifications from which the Product Structure File is created although these rules are not currently included in the P.S.F.

It may be assumed that it is mandatory to have a manual or computer substitution system in order to achieve the present negligible incidence of component shortages during assembly. If a computer system is introduced it saves £15,000.

If a proposed solution includes holding L1 and/or L2 stocks, the inevitable obsolescence-risk costs can be reduced by cutting stocks while keeping stock shortages

at whatever is considered a reasonable level, through the
use of substitutions. Such a facility should be included
in any solution involving intermediate L1 and/or L2 stocks.

11. System improvements. You are invited to criticise
the present system for dealing with orders for assemblies
at the three levels, and, while adhering to the present
delivery and assembly lead-time targets, to propose an
improved approach which will particularly help the problems of L3 stockouts and cost of L3 obsolescence, and also
of substitutions.

Please show the financial and other benefits of the new
approach, and also specify in reasonable detail how it can
be implemented on the 4G computer, showing file designs
and access methods, processing modules etc. Note any
implementation problems.

12. To help in estimating file sizes etc. the following
further information will be of assistance:

> Product Structure File (existing): this is a
> magnetic tape file consisting of three parts:
> PSF-L1, -L2, and -L3.
>
> PSF-L1 contains 250 records in assembly code
> sequence, the code is in the form 100xxx.
> Each record contains:
>
>> code: 100xxx
>>
>> descr: 30 alpha
>>
>> release date: ddmmyy
>>
>> assembly instructions: 1,000 alpha
>>
>> price data: 50 alpha
>>
>> number of components (varies 20 - 200,
>>   average 50)
>>
>> component codes in the form 0xxxxx, each
>> associated with a quantity xxxx. (Average 2)

PSF L2 and L3 are as L1 except that their assembly codes
are in the form 200xxx and 300xxx, and their component
codes will include 100xxx items, and 100xxx and 200xxx
items, respectively. Component codes are always held in
ascending sequence.

Stock File. This is used for both components and assemblies. It is a magnetic tape file in code sequence.
Stocks of assemblies are always zero for L1 and L2 at
present, although there may be several in work to meet
customers' orders. Stocks of L3s are always 1 or 1, and
there may be 0 or 1 Works Orders correspondingly.

Components show orders placed on suppliers (for progress chasing, which may be assumed to be satisfactory) and balance on hand.

Code (0xxxxx for components
      100xxx L1s
      200xxx L2s
      300xxx L32)

Balance on Hand xxxxx   Averages: components 1000
                                      L1s and L2s   0
                                      L3s   .5

Number of orders outstanding xxxx
                    Averages: components   2
      L1s $5 \times (12+5(13+25\times5))/250$ 14
      L2s $5 \times (13+25\times5)/250$ 3
      L3s 0 or 1 $\frac{1}{2}$

For each order (all types)
      order reference no. xxxxxx
      date placed   ddmmyy
      quantity   xxxx
      when expected   ddmmyy
      supplier or
        customer   xxxxxx (zero if Works Order)

Stock Control information - suppliers, lead-times, demand history etc. 200 characters.

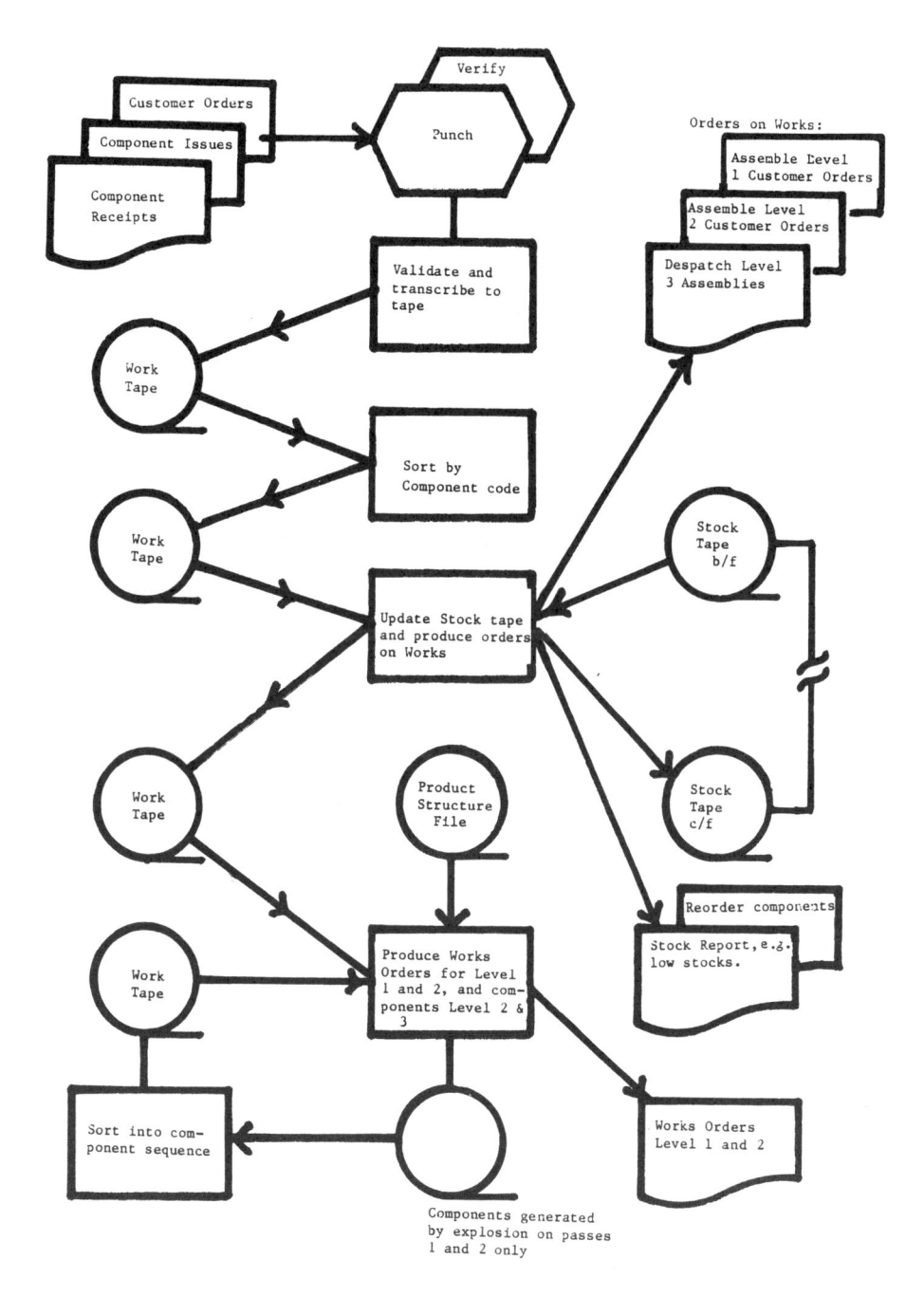

[Flowchart of production system data processing]

Inputs: Customer Orders, Component Issues, Component Receipts → Verify Punch → Validate and transcribe to tape → Work Tape → Sort by Component code → Work Tape → Update Stock tape and produce orders on Works (inputs: Stock Tape b/f; outputs: Stock Tape c/f, Orders on Works: Assemble Level 1 Customer Orders, Assemble Level 2 Customer Orders, Despatch Level 3 Assemblies, Stock Report e.g. low stocks, Reorder components) → Work Tape → Produce Works Orders for Level 1 and 2, and components Level 2 & 3 (input: Product Structure File; outputs: Works Orders Level 1 and 2, Components generated by explosion on passes 1 and 2 only) → Work Tape → Sort into component sequence

Current Production System: customer orders for Level 3 items are met from stock if possible. Other level components are assembled or ordered as a consequence of an actual customer order, so explosion system is _gross_ not _netted_ against stock at each level.

14
Hardware and Software for Case Studies

All equipment and pricing is imaginary. Purchase prices are quoted: if equipment is purchased add 10% for maintenance per annum. Equipment may also be rented at 2% of purchase price per month, with no maintenance charge. This may be useful for staged systems development.

Reliability code: (see next page)

TT 11 char/sec teletype with keyboard, tape reader/ punch, printer £1,000 A

MM Modem for TT, TV, or RT: 100 B

Telephone charges/minute:

	Normal hours	Outside normal
Under 10 miles	.2p	.1p
10 - 34	3p	1.5p
35 - 50	5p	2.5p
over 50	7p	3.5p	C

CC Concentrator. Allows up to ten lines to be carried for cost of one. Requires DC at other end. £5,000 B

TV Like TT but CRT display, no hard copy or paper tape, with switch to allow TT instead, 22 ch/sec £1,000 B

RT Remote batch terminal, 300 lines or cards per minute. (Requires MM and RT at other end, or telecomms on computer.) £15,000 A
Magnetic tape and keyboard option: MT £2,500 A
Each keyboard up to 10 KB 750 A

SG Switch gear: permits unlimited switching between lines. £1,000 C

DC Deconcentrator £5,000 B

LA Low speed line adaptor to interface to computer (up to 10 lines) £2,500 B

HA High speed for RT, otherwise as LA. One line
maximum. £2,500 B

TX Telex equipment and line attachment. £500 per year A
Add on line charges.

MX Facsimile transmission as TX A

VI Cost of installing intercom in vehicle
other costs as telephone. Range 50 £1,500 A
miles.

VH Central radio installation £5,000 A

RF Automatic telephone answering unit £750 per year A

CP Orthodox card punch/verifier. Or £3,000 A
you can simply charge cost of punching and verifying 80 characters,
including operators and punches, at 3p

Clerks in general may be charged at £1.50 per hour.

MF Microfilm/microfiche producer. Generates
one copy (equivalent of one page of
printer) per second from magnetic tape.
Also includes an off-line copier which
produces further copies of each at a
cost per frame of 1p. £20,000 A

MV Microfilm/microfiche viewer £200 C

Service Bureau: time-sharing service for TTs costs £10 per hour inclusive of processor (mill) charges, connect time, file storage: exclusive of TT, modem, line charges by GPO, and program development. Bureau has backup computer so reliability is C. TT and modem as given.

Note on reliability. Equipment is of three types,

A fails in any hour with probability .005 and is down 1 hour
B fails in any hour with probability .001 and is down 5 hours
C negligible failure rate

Mini-computer

Can drive a local TT or RT without modems and is capable of simple calculations on files read and written on 100 c.p.s. paper tape only. BASIC programming and about 16K of 16 bit words. Programs cost around £1,000 to write and test, and computer time for this can be assumed to be included. Price for 16K, 100 cps paper tape in/out, and interface for TT (but not TT itself) is £10,000. With an RT attached it can drive it at 300 cpm/lpm. You can also buy a card punch (200 cpm) for £2,500.

D.I.Y. Microprocessors

You can build your own system by buying microprocessors at the following prices:

    Add, subtract, interrupt detection, memory transfers, etc.
    8, 16 or 32 bit at 2 usec. £20, £25, £30
                              for minimum qty of 10.
    4K byte store: £25
    Other peripherals can be attached as elsewhere noted.
    Power supplies, boxes etc. have to be procured.

However, it is necessary to employ two specialist technicians at £8,000 for one year to produce a viable configuration, and there is a 50% probability of this extending to two years.

Computers

(The "4G" range) Reliability is B in all cases.

----------------COST £-------------------

Processor size:	64K	128K	256K	512K	1024K
Speed used 1	5,000	8,000	15,000		
.5		10,000	18,000	25,000	
.25		12,000	20,000	30,000	40,000
.125			25,000	35,000	45,000
.0625				40,000	50,000

(words or bytes, it doesn't matter)

Discs. 500K/sec transfer rate (so watch out for overloading CPU if several transfers occur on different channels at the same time). £10,000 per drive.

Reliability A, rotational delay 20 ms, seek time 40 ms, capacity 15m characters, but of course 100% usage is not attained with random access files.

Fixed head disc. 2M/sec transfer rate. £15,000 per drive, reliability B, rotational delay 5ms, no seek time, capacity 5m characters.

Line printer 1100 lines per minute, £7,000, reliability A.

Card reader, 1000 cpm, £5,000, reliability A.

(Note - the above devices are assumed to have their own built-in controllers).

Multiplex channel for LP, CR and Telecommunication devices. Unlimited capacity and no charge! Reliability C.

High speed channels or 'ports' for tapes and discs. Each can handle transfers from one drive at one time, but seeks can be carried out in parallel. £2,000 each, C.

Audio response unit. Accepts remote telephone as an output device: can speak digits and fifty other predefined words. However, in UK it is not possible to dial up the computer and use the dial or buttons to convey data to the computer, except where an internal line is used, in which case a sender can convey information or requests. Price £5 000, reliability B.

Computer room and airconditioning. Add 25% to the cost of equipment installed. Not required for mini- or remote batch terminals, or for micros.

Tapes £7.50 per reel. Discs £200 per pack. Fixed head discs inbuilt in drive.

Operators. Allow for one manager at £7,500 and one operator per shift at £4,000 plus one extra. Mini does not require an operator beyond the clerk who uses it.

Program development. Assume it takes six man-months per module, charged at £5,000. Each module needs the equivalent to one million instruction executions to compile and test. Alternatively use a bureau, which costs £2,000 to prepare each module after coding. Minis and micros - see in that section.

Systems design, training, cutover etc. Make allowances for this based on your own assessment. Allow for the design of fall-back/recovery, stand-by systems etc. It may be cheaper to have two small processors than one big one if the fall-back for the big machine costs more than is saved by Grosch's Law.

Costs of consumables such as stationery can be ignored - assumed to be as under the previous system.

In assessing core requirements, assume the operating system takes:

16K for running single stream of batch jobs	20%
24K to run two batch jobs	30%
4K to buffer low speed devices ("SPOOL")	2%
64K to run real-time processing excluding buffers	100%
32K for first 'MOP' and 8K for each thereafter	20% + 5% +...

To assess processor activity taken up by the operating system, compute the application program and I/O requirements in terms of cycles per program module (as shown below) and add on overhead percentage above.

Application program requirements. Assume each major transaction type is handled by one program module of 24K core size, requiring 100K processor cycles to complete plus I/O transfers at one cycle per character.

In batch as distinct from transaction-oriented systems designs, still use 24K programs but estimate the number of runs in the suite in the usual way, not omitting sorts.

Hence, to run 10 transactions (5 different types) a second in real time, and avoiding the need for overlays, will require 5 x 24K plus 64K plus buffer areas for disc etc., and the cycles required per second will be 10 x 100K per second thus requiring probably .25 usec core, which actually provides 4,000,000 cycles.

It should however be noted that disc transfers may be the limiting factor, or the size of buffers available for queues and records.

Buffers. Allow 1K for every file, x the number of transactions in process together, if real time. If batch, double-buffer any serial files. Hence, in the last example, if in the most complex transaction type four files are open in the application program for each transaction, when 10 transactions are in process together up to 40K of disc buffers are required.

Buffers should also be allowed for messages going to and from lines, according to message length.

Tapes - A 200K/sec unit is available for £5,000: reliability A. Stop/start time 10 ms. One reel holds 30m characters (at 1K blocks).

Author's Notes on the Case Studies

These notes are in no sense model answers. They may well be that people tackling the case studies may even find approaches which the author overlooked and which may well be better than his. These are some of the ideas which entered his mind when writing and later on, administering the case studies. I owe a debt to several generations of fourth year Brunel computer science students for their contributions as well.

Minim's Restaurant

This case study is rather like a detective story. The really significant fact is that an enormous amount of money is being lost. The raw materials should cost £161 980, i.e. 35% of the revenue £462 800. But the actual cost are stated at £250 000. This means that £88 020 is being lost through inefficiency or fraud. It is not clear from the information given how this loss should be distributed between inefficiency and fraud, but further analysis of the procedures will show that it is possible for most of the staff to misappropriate money or food or drink. This may even be done with the connivance of the management, as a way of giving a tax-free supplement to the low wages of the staff: as in Agatha Christie's Murder on the Orient Express, perhaps everybody is guilty.

The systems analyst in this situation would be foolish to introduce computer systems which ignore the fact that the enterprise is being run on lines quite alien to those assumed by UFO. If his new system turns out to be fraud-proof, the staff will do their best to sabotage it. If it is not fraud-proof, Minim's will enjoy computer-assisted embezzlement procedures.

Seeing that the students in some cases were ignoring this evidence, the author distributed a copy of what was supposed to be a letter from the Japanese Managing

Director of UFO's pearl fishing enterprise, talking about his own problems, but also mentioning that he had a good dinner at Minim's Restaurant, when passing through London, but quoting the price at around £40. This document was meant to alert the students to the possiblity that something queer is going on at Minim's. Oddly enough, many students chose to overlook this and pressed on with a conventional computer systems report recommending the installation of hardware and software.

It seems that the proper course of action for an analyst would be to prepare a report (which is marked confidential) for top management in UFO: it should be pointed out that there is evidence of gross inefficiency and/or fraud, which must be investigated by auditors as soon as possible and that the likely sequence of events will be: the replacement of Minim's management (it does not really matter whether it was inefficiency or fraud which caused the losses - in either case management is to blame), the installation of some procedures to identify exactly where the losses are occurring, possibly using sampling techniques, and only finally the introduction of sophisticated computer systems. We may imagine that UFO will be faced with a difficult industrial relations problem: if they intervene in the operation of Minim's in a way which prevents inefficiency and fraud, they may be faced with a mass walk-out of staff. They may then have to back down, or else change Minim's style from a traditional London restaurant to something closer to Kentucky Fried Chicken or MacDonald Hamburgers. UFO might just conceivably decide to let things go on as they are.

The computer science student may well ask, what then is the pure computer science content of this case study? Firstly, it underlines the computer professional's responsibility to his employer to point out situations where something other than the introduction of a computer system is the first priority. In addition, part of the tool-kit of a computer professional should be the ability to analyse clerical procedures properly using such techniques as flow charts and decision tables (not to mention good English narrative) and so be able to describe manual systems and their shortcomings. He should also be something of an accountant and be able at least to do the simple analysis noted early. In the author's view he or she ought also to go a step further and in the second half of their report describe how eventually, computer systems may be used at Minim's, on the strict understanding that the problems have been sorted out. If we read between lines, it is clear that UFO are keen to make use of their "Magic Star" reader and to couple Minim's to their central computer at Phoenix. Some quite nice systems can be designed which will help the kitchen plan its production schedules, keep track of raw materials, and enable the waiters to take orders. For example, the menus can be produced in such a way that each item on the menu has a Magic Star beside it, and the waiter by touching the appropriate Magic Star with a light pen can indicate which items on the menu are being ordered by the customers. One other area perhaps deserves a note: the booking of tables.

Clearly the head waiter has an interest in preserving the current system but it may well be that it actually reduces the profits of Minim's by reducing utilization of tables. The report should include some suggestions as to how analysis, or, if necessary, an experiment, could find out if this is the case. If booking does seem to be a good thing, the computer system should be able to include it.

In summary, this case study is an opportunity for the student to try out his skills in procedure documentation, writing a tricky political report, and designing and costing a real time order entry system. At all times he has to think about the motives of all the people involved in this situation: his masters at UFO headquarters, the management and staff at Minim's and last but not least, his own career objectives. Minim's may now be a bucket of worms but could turn out in the long run to be a feather in his cap!

The Dental Supply Company

This is another case study with a rather large 'human factors' element in it. It is possible to use it in two ways, either the team can role-play the Rationalistics systems team and design the on-line order entry system or it can role-play the consultant team which tries to sort out the mess. In the latter case the outline systems design is part of their brief. Since the author has given a possible 'systems' solution to the case in the text, no further comment is needed here. However if the team adopts the role of management consultant, the following points may be helpful.

Although it is true that the computer system coupled with

the integration of the two companies will lead to many benefits, it is also true that a good proportion of these benefits could be gained by tightening up the very sluggish procedures at DSC. Much too long is spent in processing an order and the cost of this control seems quite out of proportion to the savings of bad debts that it achieves. In addition of course we can see that the introduction of the computer system, as in the Minim's case, is going to be fraught with personnel problems: the existing DP department is set against such a development, and the change in role of the current order clerks at DSC is going to be industrially unacceptable. One might note that the brief mentions in passing that the Production Director has stated that improvements in his control procedures could cut costs by up to 5%. This is on an annual cost of £20 000 000, in other words if he is right, the operating surplus could be increased by 50% through improvements in this area alone.

This case study has been presented at several courses given by the Management Programme at the Brunel Institute of Social Science to computer professionals from industry and government. It was used largely to alert them to identify who the critical people in the problem are, what their motivations are, and what their power base is. The teams were encouraged to think how negotiations with these various groups could be carried out and lead to a successful conclusion which would satisfy Tom White's objectives. Some computer professionals looking at this case have found a way of "getting Tom White off the hook". Much of the problem has arisen through the proposal to integrate, physically, the two very different order offices of DSC and OCC. Through the use of a computer fitted with communications devices and remote terminals, it would be possible to offer Tom White a system which allows for the logical integration of his systems while still keeping order clerks at the original sites.

This would give him considerable strength in negotiation: if necessary he could switch order taking from one location to the other. To some extent the author's colleagues at the Management Programme consider this escape mechanism something of a cheat, in that it ducks the issue of negotiating with all the various power groups involved. The author's view is that Tom White should indeed attempt to solve the human factors as posed, but that the technical resource offered by a computer with remote terminals will strengthen his position. The author knows of a situation exactly parallel to this where the distance between the two groups of clerks was the width of the Atlantic, and the availability of remote processing is a card that management played in industrial negotiations to enormous effect.

One or two minor points: it seems inevitable that
Mr. Johnson of the existing DP department will have to go.
OCC must be kept running, and no improvement to DSC procedures should be allowed to cause us problems at OCC: the
sums of money involved in a shut down at OCC are far
greater than any benefits of systems improvement. Probably
the best line for Tom White will be to climb down and postpone integration, saving his face by blaming the GPO for
being slow in providing telephone lines (an excuse which
everybody will believe), or possibly the Rationalistics
proposal (for being unduly optimistic in its time scale,
which it is). He can then proceed on two fronts: tightening up procedures at DSC order taken and reorganizing
the DP department and in parallel at OCC, installing an
on-line order entry system which the OCC clerks will probably accept since it will not involve them in moving. He
can then install terminals at Hammersmith for DSC use on
the fast moving items, and if the clerks object, he is in
a position to re-route such orders to the OCC order-taking
centre.

Bluebird Aircraft

This is a case study without a "human factors" dimension -
or at least, the author is not aware of one. It is a
classic problem of constructing a bill of materials
system: quite a complicated one, because it has to include
provision for alternative parts, modifications and complex
interrelations between parts. No information is given on
the staff required to run the system at present, and if
the administrator of the case study wishes he may issue a
bulletin to the syndicates giving them his own views. In
fact the benefits are given in the case as being in two
main areas: improving the number of aircraft sold by a
factor of 30% through faster delivery and a once and for all
all reduction in work-in-progress stocks. The possible
annual benefits should be worked out on this basis and
any computer system must clearly cost less to operate per
annum (including rentals or depreciation) then the benefits. It is likely that in a smallish company, which is

what Bluebird appears to be, it will be uneconomic for the software to be developed in-house, to cover the very complex requirements of this production control application, and it is more likely that they would buy a large mini-computer with database software such provided by a number of manufacturers now. However, it is not sufficient for the team simply to say that the database will provide all the facilities which are required: it must indicate what kind of a system of pointers, indices, and accessing modes will be necessary. The teams should be familiar before starting this case study with the main principles of production control systems, particularly 'Bill of Material' (or 'Assembly Breakdown') files, parts files, etc., used in the explosion of products, level by level, into components, and the planning of net requirements of components.

The team will need to modify this basic approach to cope with the special needs of this application: modification and substitution control, and C of G/weight calculation.

It should also think about the data collection problem and how the system is to be introduced.

The Garden Ornaments Management Information System Project

The computer professional is so inured to scepticism that when a potential user of his systems is over-enthusiastic, he finds it hard to restrain this enthusiasm, especially when the man in question is the chairman. It happens in some big organisations with a "grand old man" at the top, that there wells up in him a sudden passion for the latest in technology or management science, very often much at variance with the dour, traditionalist attitudes of his managers. In some cases this passion promotes strife within management, and strengthens the position of the grand old man: one could call this manoevre the 'Mao Tse Tung' tactic: although it is often hard to say whether his professed enthusiasm for innovation is assumed only, or genuine, and only inadvertently of political value to the grand old man.

I wrote this case study not long after all too relevant personal experiences, and concluded the writing of it on a note which I made as optimistic as possible: "perhaps GOMIS could succeed" - perhaps a simpler, watered-down version would be safer. But looking at this case again I feel that not much optimism is justified. The grand old man is not interested in watered-down versions, and the poor DP manager feels as if he has been caught in a rowing boat in the middle of the battle of Jutland, as the big bosses line up their guns.

It is in cases like this that the DP professional faces a real moral dilemma. If he decides wrongly, his colleagues, his staff, the company's shareholders and his own family could suffer. He may think that it may be best to play along with the grand old man's schemes, knowing them to be impractical - but if there is a palace revolution he would be thrown out along with the grand old man himself. But if he argues, his ejection might come even sooner.

One line is perhaps to accept GOMIS, but introduce it in phases, of which the first will be simple, effective and indefinitely prolonged. If the DP manager does not bale out, he should at all costs avoid politics and do the best professional job he can.

(This case study was first published in Management Decisions, 1969)

Brighton Rock

The present procedures are crude and would suffice only if the company has a monopoly: customers must frequently lose their tempers when their orders are arbitarily cancelled, and would buy their rock somewhere else if they could.

Two strategies are immediately attractive:

> To improve the FIFO scheduling algorithm so as to minimise changeovers;
>
> To have longer runs - so that demand is

satisfied from stocks instead of directly by production.

We could change the present scheduling algorithm (do low-numbered orders first) to doing the ones with earlier due dates first. This would be a simple change which would not require any computer system. However the argument against this is that the prudent customer who orders well in advance will have no advantage over the customer who orders for delivery yesterday: in addition, customers will learn to quote very early delivery dates in order to get priority.

A more thoughtful approach is for the order office to assign <u>priority numbers</u> to the orders, which are determined by the due date, order date, and the importance of the customer, and for the orders to be carried out in this priority number sequence, except that outstanding orders are to have their priority increased as time goes by.

However, neither of these approaches takes account of the advantage to be gained by selecting an order for production, which demands the minimum changeover time. So the second approach may be modified as follows: let us use a sufficiently small number of priority numbers to ensure that at any time there are five or six orders with the same priority, and when the machine becomes free, assign that order which needs least set-up. The advantages of this can be considerable.

Currently 15 x 70 minutes a week out of a possible 15 x 320 minutes (= 5 x 16 hours), are apparently spent in set-ups - nearly 22%, so there is great scope for saving time in which to make more rock.

A good way to begin to estimate this saving is to prepare a table showing the basic changeover-times for each pair of products, i.e.

		TO			
		A1	B1	C1	D1..
FROM	A1	0	15	15	45..
	B1	15	0	15	45..
	C1	15	15	0	45..
	D1	45	45	45	0..

(This assumes that hoppers can be preloaded during two-colour runs, ready for a three-colour and ignores the hopper-pump connection changes.)

It will be observed that the task is symmetrical about the diagonal.

It will be rare for us to be able to run on two orders of exactly the same kind, but it should be possible to make most changes of 15 minutes only rather than 45.

(At this point we may well question the 70 minute set-up time quoted in the brief: even if the changes between products is quite random, the average time taken should be much less than 70 minutes. Accordingly the report should urge an investigation into this: there may be a simple procedural change which could reduce the set-up time to the average, at least.)

We may now consider more sophisticated algorithms for optimising the production sequence: with one machine there are known mathematical solutions (such as dynamic programming) which will minimise the total cost of a production sequence, defined as a function of set-up times and of over-running priorities. (With several machines, the computation gets out of hand.)

But we should note the following points:

> If the decision to put a particular order onto the machine is taken by a computer rather than a human supervisor, who carries the can when an important customer's order is delayed? The systems analyst must make sure that a re-shuffle of responsibilities is going to be accepted by those concerned.
>
> The 'optimising' algorithm may grind away at many complex equations, but one vital ingredient - the priority number - is a very rough and ready, subjective figure anyway. It is ridiculous to use sophisticated algorithms on data of low precision.
>
> A reasonable compromise could be to install a microcomputer which holds the table of change-over times and the outstanding order queue, with their manually assigned priorities.
>
> At any time it will arrange the orders in the highest priority classification into a sequence which, following the order just being run, will minimise set-ups. But it will never run a low priority job ahead of a high priority.

The question arises, would the microcomputer pay its way?

The best way to answer this is to run a _simulation_, and in fact when my students at Brunel University were faced with this case, some of the syndicates wrote and ran small evaluation programmes to see whether, on a random mix of orders, the microcomputer would achieve a reduction in set-up times which more than paid for its cost (£2 000 or so) and was _significantly_ better than the manual system suggested earlier (load _any_ order in the high priority group which has minimum set-up time, and do not look ahead to see whether this may cause trouble with later orders.)

Now we come to the more radical approach - instead of trying to improve the scheduling of production-to-order, why not make for stock, and enjoy really long runs: if we double the length of a run, we can halve the proportion of time spent in set-ups. But there are problems:

> We need to check there is room to store the rock, and that it does not deteriorate on the shelf. Brighton Rock staff will need to be trained in the special procedures for holding, allocating, releasing, checking and costing finished stock.
>
> There will be a problem of accumulating enough rock to introduce the scheme: if we switch over to long runs straight away, our existing orders will be met even worse than before. Perhaps we should introduce a third shift of weekend working. (In fact, why not do this now anyway, since there is unsatisfied demand?)
>
> We need to calculate how big the standard-batch-to-make (and by implication, the average stock level) of each product ought to be. This is the good old EOQ calculation.
>
> The computer system, if used, will now be concerned not so much with optimising production as administering stocks, filling orders, and, no doubt, invoicing and sales ledger. We may need to go back and suggest new terms of reference for our study.

Camelot Ice Cream

This is a rather massive, multi-dimensional case which requires the syndicate to lay out its effort carefully. A good first step is to complete the quantification of benefits in the section entitled "Opportunities and Problems", and to concentrate on a few high potential value areas.

A second step should be to draw the present procedures in the form of a _columnar flowchart_ showing how information - and ice-cream - flows from location to location, with boxes representing procedures carried out, cross-referenced to narratives, _decision tables_, and/or _logic block diagrams_ elsewhere.

The team is then in a good position to identify which targets to aim at, and which parts of the procedures are crucially affected. It may be worthwhile putting up the 'present system' chart on the wall of an office, and for the syndicate members to spend time in front of it, pointing at it and arguing.

It would appear that there are two main lines of approach: one is to start at the centre and refine clerical procedures in the stock, accounting, and central distribution areas: the other (more dangerous but potentially more profitable) is to tackle the difficulties at the point of sale, which will perhaps involve proposals for new methods of paying salesmen which should both increase their own earnings, and Camelot's profits, by motivating them to make more sales, fewer returns, and less undercover resale of time-expired packs.

Or of course you can propose both, and choose which to do first on the basis of logical dependence (if any), and rapidity of pay-off.

Vagabond Car Hire

This case study has a strong flavour of Operational Research, and an appropriate response from a team of systems analysts would be a suggestion that Operational Research specialists should be invited to conduct an analysis of the proposed system, probably using queueing theory and/or simulation. (It must not be assumed that, because Hertz and Avis can operate a "Wanderlust" scheme, so can a much smaller organisation because with so few cars Vagabond is at the mercy of statistical fluctuations which could leave all its cars at one location.)

The computer professional can contribute the design (and costing) of a suitable control system to support Wanderlust: probably microcomputers with floppy discs at each office for recording hirings and returns, and producing hire forms: the information on the floppy discs can be posted, sent by courier, or transmitted over GPO line to a central point. Here it should be possible to foresee and forestall shortages of cars at particular locations, collect information on cash receipts etc..

The costs and security of such a computer based approach should be compared with a reasonable manual system - e.g. using Telex.

The teams may also consider whether there is an argument for installing such a system, whether or not Wanderlust is fully implemented, on the grounds that it will improve security, reduce costs, and increase utilisation even under present rules, and that a controlled experiment could be carried out by allowing a limited number of offices to operate Wanderlust on a pilot basis. (There are arguments, in fact, for mounting such a trial rather than just doing a paper or computer simulation: we need to know the actual reaction of real customers as well as the incidence of imbalances in vehicle stocks.)

Canine Eugenics

There is in fact a company in the USA which offers a "dogs dating service", although for the entertainment of the animals rather than the profit of the owners.

A little analysis will show that the proprietor's claim to carry out 'optimum' matching, according to his "magic formula" is unlikely to be true. The team must therefore try to discover just how far a matching process ought to go to be considered satisfactory. One can, for example, construct a scheme as follows: for each·breed, set up one file of dogs and one of bitches. Read a number n (about 10, say) from each into a pair of buffers in computer memory. Find the best-matched of the 10^2 = 100 possible pairs, assign them together, then replace their entries in the buffers with another pair of records, one from each file. To be fair we ought to shuffle the dog and bitch files occasionally since the nearer the beginning of a file an animal is, the more chance it has of a good match. Enthusiastic syndicates can run a small simulation to discover the best buffer size to balance computer time (exponentially increasing with buffer size) and excellence of match (increasing more and more slowly with buffer size).

This is obviously suboptimum - if the first ten dogs are matched ideally with the last ten bitches, the system will not discover it - but then it may well be adequate for marketing purposes.

We may well query the whole reason for introducing the computer system - it may just be a gimmick to boost sales rather than a genuine attempt to improve performance. We need to talk frankly to the proprietor. However, the straightforward book-keeping task is probably well worth putting on a small computer. The mention of mark-sensing may lead a team to design an application form which can be processed without conventional key punching.

Banzai Pumps

This is a case concerned with file design, and in particular the use of computers to support marketing management: the construction of an extended customer file which can represent the structure of, particularly, the 'special' customers, and their contractual arrangements.

The programs will be required to traverse these complex customer structures when reporting on, for example, the history of complaints or of discounts negotiated - this means that a database of some complexity has to be designed, or else procured (e.g. TOTAL, ADABAS) and adapted. It will be noted that only a minority of customers warrant this complexity, and the teams will have to think whether to split the customer file into a simple one, of 'normal' customers, and a smaller, more complex one of 'specials'. Processing will be quicker in the latter case, but the complications of handling two file types may be more expensive in systems effort than the processing time saved.

A further problem is the handling of old records, now held "indefinitely". The systems survey should consider the regular purging of records to magnetic tape and thence to microfilm/fiche or possibly holographic storage.

Syndicates should have been prepared for this case by lectures on file design, particularly the use of index-sequential files, and chained structures.

The PNP Electronics Company

This case study is concerned with <u>product structure</u> (or "assembly breakdown" or "Bill of Material") files, with particular emphasis on the problems - and opportunities - presented by components used on several products.

As a hint to those tackling this case, companies whose products do have many subassemblies in common will often make such subassemblies to form <u>part-finished</u> stocks, from which finished products ready for sale can be assembled quite quickly when ordered. This approach is a good compromise between the expense of keeping a large stock of finished products, which are likely to become obsolete, and represent a heavy investment of cash, and the expense of protracted manufacturing lead-times (in terms of dissatisfied customers) and fluctuating loads on men and machines.

BIBLIOGRAPHY

Diebold Group (editors), *Automatic Data Processing Handbook*, McGraw-Hill, New York, 1977

Hartman, Matthes and Proeme, *Information Systems Handbook*, Kluwer-Harrap, London, 1968

Judd, D. R., *Use of Files*, Macdonald/American Elsevier, London, 1973

Price, W. T., *Introduction to Computer Data Processing*, The Dryden Press, Illinois, 1977

Purchall, F. W., and Walker, R.S., *Case Studies in Business Data Processing*, Macmillan Press, London, 1972

Race, J. P. A., *Computer-based Systems*, Hodder and Stoughton, London, 1977

Radford, J. D., and Richardson, D. B., *The Management of Manufacturing Systems*, Macmillan Press, London, 1977